Edition XII

*worldwide directory
of postgraduate
studies in*

Engineering
and
Technology

1997/98

Publisher
Edward More O'Ferrall

Production Director
Gary Walsh

Production
Christine Martinez
(Project Coordinator)
William Booth
Pilar Blein
Christopher Clark
Barbara Icardi
Janet Josephs
Marie McGonagle
Campbell Reid

Internet Development
Simon Dayton

Director of Marketing
Jan Field

Sales Manager
Paul Marshall

Sales
Martin Brophy
Anthony Butler
Max Clapham
Inès Demonchy de la Chaise
Morgan Grover
Niall Kennedy
Damien Lesavre
Manuel Navarra
David Newman
Richard Ramsay
Ramsey Taymani
Elize Truter
Dr Omar Wahab

Edition XII Limited

10 Regents Wharf All Saints Street London N1 9RL United Kingdom

Second edition

© Edition XII Limited 1996

ISBN 1 86149 009 7
ISSN 1361-343X

Printed by The Burlington Press, Cambridge

Publisher's note
The information contained in this book has been published in good faith on the basis of information supplied by the institutions listed. Edition XII Ltd cannot guarantee the accuracy of the information in this book and accept no responsibility for any error or misrepresentation.

Contents
General

Contents
by Country

Contents
by Country and Programme Category

CONTENTS BY COUNTRY AND PROGRAMME CATEGORY

South Africa

Engineering

General Engineering

Management

Spain

Communication Engineering

Electronic Engineering

General Engineering

Mechanical Engineering

Sweden

Ukraine

United Kingdom

CONTENTS BY COUNTRY AND PROGRAMME CATEGORY

Mechanical Engineering

Nuclear Engineering

on the net **http://www.editionxii.co.uk**

Contents
Extended Entries by Country and Institution

The Edition XII Series of Publications

The Edition XII worldwide directories of postgraduate studies are the most universal directories on postgraduate educational opportunities on the market.

Aiming both at those planning to study at home in their own country, and at those intending to study abroad, the directories feature profiles from a wide variety of educational institutions worldwide. Each institution is represented by concise information on courses offered, departmental/institutional resources, staff profiles and information on dates, fees etc. The directories build up into an attractive series presenting an unrivalled cross-section of postgraduate education opportunities.

The following titles are available:

Edition XII worldwide directory of postgraduate studies in:

* **Arts, Humanities and Social Sciences**

* **Business, Economics, Finance and Commerce**

* **Engineering and Technology**

* **Science**

* **Computer Science, Mathematics and Information Technology**

* **The Directory of MBAs**

Edition XII also publishes *The Directory of MBAs*, the leading authority on MBA programmes. MBA opportunities from more than 35 countries are profiled in an attractive, informative, full-colour format. The reference section comprises short user-friendly entries from more than 50 countries, giving the broadest overview of MBA programmes available today.

All Edition XII publications appear on the Internet, at the Edition XII web address, *http://www.editionxii.co.uk*, one of the fastest growing educational sites. Each entry appears on their own page, together with hypertext links direct to their own web site. A full postgraduate reference section is also available at the Edition XII home page.

Coming soon for 1997:

* **Open Forum: The Directory of Open Learning**

THE NEW FRENCH DIALLING CODES

From 18 October 1996, 23:00 French time (21:00 UTC), the changes outlined below will come into effect in France. As of this date, all numbers in France will consist of a nine-digit number after the International Dialling Code (0033 or +33, as used in this *Directory*).

Please note that information on courses and institutions in France represented in this *Directory* was supplied before the new system come into effect.

All information supplied by France Telecom (UK), August 1996.

PARIS AND ITS ENVIRONS

There are no changes here; dial the usual nine-digit number, starting with 1.

THE FOLLOWING NUMBERS REMAIN THE SAME		
1 30	1 43	1 49
1 34	1 44	1 53
1 39	1 45	1 55
1 40	1 46	1 60
1 41	1 47	1 64
1 42	1 48	1 69

OUTSIDE THE PARIS REGION

Dial either 2, 3, 4 or 5, depending upon the region, before the existing eight-digit number.
For example: (+33) 32 becomes (+33) <u>2</u> 32
(+33) 71 becomes (+33) <u>4</u> 71

THE FOLLOWING CODES SHOULD BE USED WITH NUMBERS STARTING AS BELOW			
3 20	2 41	5 61	3 81
3 21	4 42	5 62	3 82
3 22	2 43	5 63	3 83
3 23	3 44	5 65	3 84
3 24	5 45	4 66	3 85
3 25	5 46	4 67	3 86
3 26	2 47	4 68	3 87
3 27	2 48	4 69	3 88
3 28	5 49	4 70	3 89
3 29	4 50	4 71	4 90
2 31	2 51	4 72	4 91
2 32	5 53	4 73	4 92
2 33	2 54	4 74	4 93
5 34	5 55	4 75	4 94
2 35	5 56	4 76	4 95
2 37	5 57	4 77	2 96
2 38	5 58	4 78	2 97
2 39	5 59	4 79	2 98
2 40	3 60	3 80	2 99

SPECIFIC SERVICES

(For example, Videotex or Audiotex).
Dial 8 before the existing number; the number is otherwise unchanged.
ie. (+33) <u>8</u> etc

MOBILE SERVICES

Dial 6 before the existing number; the number is otherwise unchanged.
For example: (+33) 01 becomes (+33) <u>6</u> 01

THE FOLLOWING CODES SHOULD BE USED WITH NUMBERS STARTING AS BELOW		
6 01	8 04	6 08
6 02	6 05	6 09
6 03	6 07	8 36

THE OVERSEAS 'DEPARTEMENTS' AND TERRITORIES

All subscriber numbers remain unchanged. Dial the correct three-digit international code for French overseas 'départements' and territories, followed by the six-digit number.

DEPARTEMENTS/TERRITORIES AND THEIR INTERNATIONAL CODES	
Guadeloupe	590
Guiana	594
Martinique	596
La Réunion	262
St Pierre et Miquelon	508
Mayotte	269
New Caledonia	687
French Polynesia	689
Wallis and Futuna	681

Royal Melbourne Institute of Technology
Department of Aerospace Engineering

Taught Courses
- Bachelor of Engineering (Aerospace)
- Bachelor of Engineering (Aerospace) / Bachelor of Business (Business Administration)
- Bachelor of Applied Science (Aviation)
- Bachelor of Applied Science (Aviation) / Bachelor of Engineering (Aerospace)
- Associate Diploma of Engineering (Aerospace Systems)
- Master of Engineering by coursework

Research Programmes
- Master of Engineering
- Doctor of Philosophy

MAJOR RESEARCH AREAS
- Advanced composite structures
- Guidance, control and dynamics
- Aerodynamic loads
- Computational mechanics
- Crashworthiness
- Aircraft design

General departmental Information
- Located at the heart of the Australian aerospace industry, at an industry site
- Largest aerospace teaching and research programs in Australia
- Excellent laboratory and computational facilities
- Conducts joint research with Department of Mathematics
- Attracts top quality undergraduate and postgraduate students

Outstanding Achievements of the Academic Staff to Date
- World-ranking research in textile composites and optimal control
- Designed and produced an advanced composite monocoque racing cycle frame for the Australian national team
- Staff are regular winners of excellence and achievement awards

Special Departmental resources and Programmes
- Research centre: The Sir Lawrence Wackett Centre for Aerospace Design Technology
- Approved by the Civil Aviation Safety Authority as an authorised aircraft design organisation
- A centre of expertise for aircraft flight loads
- Leaders in textile composites engineering
- A major partner in the national Cooperative Research Centre for Aerospace Structures
- Strategic alliances with major industry partners

Outstanding Local Facilities and Features
Melbourne has been voted as the "most liveable city in the world" in a recent international survey.

With a cosmopolitan population of over three million, it is a major cultural, sporting and technology centre.

Contact
Professor Lincoln A Wood
Department of Aerospace Engineering
Royal Melbourne Institute of Technology
GPO Box 2476V
Melbourne Vic 3001
Australia

Tel (+61) 3 9647 3098
Fax (+61) 3 9647 3099
Email
wackett_centre@rmit.edu.au
WWW
http://www.aero.rmit.edu.au

Deferred payment option
AU $14,500–AU $16,000
27,231
1 : 20
20%
Some rsrch schlshps av
Yes
Feb 1997–July 1998
Very good
Excellent
Industry site
AU $100 per week
March 1977
25 km
State capital
State capital

James Cook University of North Queensland
Chemistry and Chemical Engineering

Contact
Professor Robin Bendall
Department of Chemistry and
Chemical Engineering
James Cook University
Townsville
Queensland
4811
Australia

Tel   (+61) 7781 4343
Fax   (+61) 7725 1394
Email Robin.Bendall@jcu.edu.au
WWW http://www.jcu.edu.au/dept/
Chemical_Engineering_and_Industrial

Taught Courses

BACHELOR OF ENGINEERING BENG (CHEM. ENG.) (FOUR YEARS)

BACHELOR OF SCIENCE BSc (IND CHEM.) (THREE YEARS)

BACHELOR OF SCIENCE BSc HONOURS (IND CHEM.) (ONE YEAR)

BACHELOR OF APPLIED SCIENCE BAPPSC (THREE YEARS)

BACHELOR OF APPLIED SCIENCE BAPPSC HONOURS (ONE YEAR)

Research Programmes
The Division of Chemical Engineering has an excellent research record. Applicants are encouraged to apply for research in the following areas: Extraction of valuable products from sugar cane; HAZOP (Occupational health and safety); Corrosion behaviour; Minerals processing; Passive fire protection systems; Electrorefining of copper and mass transfer.

MSc and PhD programmes are offered.

General Departmental Information
The Division of Chemical Engineering is only five years old at JCU. The Division's staff are highly qualified and enthusiastic with a great vision for development of the Division.

Outstanding Achievements of the Academic Staff to Date
Large research grants – ARC in sugar technology and minerals processing. Substantial industrial support and collaborative research programs.

Special Departmental Resources and Programmes
The training in Chemical Engineering is very "hands-on". The classes are small and allow or more personalised tuition. There is a full scale pilot plant on campus and a well equipped engineering bay. Modern computer design programs are also available.

The Division enjoys good working relations with local industry eg in minerals (copper, nickel and gold) and sugar.

Outstanding Local Facilities and Features
The university is located in the capital of north Queensland, Townsville. The city has a population of +/- 130,000 people and is in the heart of the Great Barrier Reef. This is great for snorkelling and scuba diving. The climate is tropical allowing a varied outdoor lifestyle. Close to Townsville are World Heritage Rainforest areas. The city offers many facilities with museums, aquarium, libraries and shopping malls.

🏫	AU $14,000 pa for 4 years
$¥	–
👥	28
📊	3.6%
%	30% postgraduates
✋	Yes
🏅	–
🗓	November 1997
📖	Excellent
💻	Excellent
?	Large open country sting
🛏	AU $140
📅	February 1997
✈	10 km (in Townsville)
📷	–
🚗	–

The University of Queensland

THE UNIVERSITY
OF QUEENSLAND

AUSTRALIA

Taught Courses
- PhD
- Master of Engineering Science
- Master of Engineering Studies
- Master of Environmental Management
- Master of Engineering Technology Management

Research Programs
- Adsorption and Chemical Engineering Science
- Biotechnology
- Environmental Management
- Particle Technology
- Plastics and Materials Rheology
- Process Engineering

All of the above areas of research are well-resourced and equipped and the department can offer students excellent supervision and support.

General Departmental Information
The Department is recognised as a leader in the chemical engineering field and enjoys a strong rapport with industry. It is involved with four national co-operative research centres in the areas of Waste Management and Pollution Control, Mining Technology and Equipment, Black Coal Utilisation and International Food Manufacturing and Packaging. It has extensive links with research centres within the University including the Advanced Wastewater Management Centre, the School of Marine Science, the Centre for Molecular & Cellular Biology, the Technology Management Centre, the Centre for Microscopy & Microanalysis and the Centre for Mined Land Rehabilitation.

The Department currently has over AU$1.5 million contract research projects for more than 30 clients in a range of industry sectors. It is very well supported in competitive Federal research grants.

Special Departmental Resources
The Department maintains a category 1 analytical laboratory which also provides consultancy testing to industrial clients. Other facilities include an electronics workshop, a mechanical workshop, excellent computing facilities including access to the University's central supercomputers. Staff from the Department have access to the University's Low Isles and Heron Island Research Stations on the Great Barrier Reef.

The Department is also active in the area of continuing professional education and co-ordinates the annual International Winter Environmental School which now attracts over 400 delegates world-wide.

Outstanding Local Facilities and Features
The University of Queensland is located in an ox-bow of the Brisbane River, about six kms from the Centre of Brisbane City (population approximately 1,000,000).

Brisbane's magnificent semi-tropical climate allows for year-round enjoyment of sports of every type and the campus has a host of sporting and leisure facilities. Its proximity to coastal beaches which are easily accessible by electric train and bus allows enjoyment of surfing, sailing and swimming activities.

The arts are serviced by a cultural centre comprising theatres, art galleries, museum as well as the State Library, a short walk from the City Centre on the former site of Expo 88.

Contact
Maryanne Glynn
The University of Queensland
Queensland
4072
Australia

Tel   (+61) 7 3365 3708
Fax   (+61) 7 3365 4199
Email
maryg@cheque.uq.oz.au.edu
WWW
http://www.cheque.uq.oz.au.edu

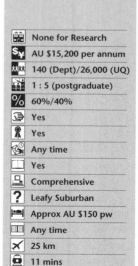

None for Research	
AU $15,200 per annum	
140 (Dept)/26,000 (UQ)	
1 : 5 (postgraduate)	
60%/40%	
Yes	
Yes	
Any time	
Yes	
Comprehensive	
Leafy Suburban	
Approx AU $150 pw	
Any time	
25 km	
11 mins	
6 km	

THE UNIVERSITY OF QUEENSLAND

The University of Queensland
Civil Engineering

Contact
Douglas J Shaw
Department of Civil Engineering
The University of Queensland
St Lucia 4072
Brisbane
Queensland
Australia

Tel (+61) 7 3365 3874
Fax (+61) 7 3365 4599
Email shaw@civil.uq.edu.au
WWW
http://www.uq.edu.au/civeng/
civil.html

Taught Courses
– Doctor of Philosophy (three years)
– Master of Engineering Science (Research) (one to one and a half years)
– Master of Engineering Studies (Coursework) (one to one and a half years)

Research Programs
– Fluid mechanics and hydraulics, including wind engineering, environmental modelling, coastal and estaurine processes, and groundwater transport modelling
– Geomechanics, including mine-site rehabilitation
– Structures and solid mechanics, including concrete structures, computer modelling and structural analysis and behaviour

General Departmental Information
The department has 16 full-time academic staff and six research staff, with a strong postgraduate program enrolling more than 50 students in research degrees.

Outstanding Achievements of Academic Staff to Date
Staff members in the Department have been recipients of national and international prizes and fellowships; for example prizes for best technical papers have been won recently by Dr P F Dux, (American Society of Civil Engineers), Dr D K Brady (Institution of Engineers, Australia).

Recent Fellowships have been awarded to Dr D J Williams, (Australian Minerals & Energy Environment Foundation), Dr H Chanson, (Australian Academy of Science). Involvement in Research Centres – includes the Department's Centre for Transmission Line Structures (Director, Professor S Kitipornchai); staff also have substantial involvement in other centres, e.g. Co-operative Research Centre for Ecological Sustainability of the Great Barrier Reef (Professor R E Volker), Centre for Mining & Environmental Rehabilitation (Dr D J Williams, Dr D A Lockington).

Special Departmental Resources and Programs
There are four large well-equipped laboratories, housed in the same complex, for experimental research in concrete, geomechanics, hydraulics, and structures. Major research facilities include a computer-controlled Instron testing system on a strong floor, a large boundary layer wind tunnel, a laser-doppler velocity meter, and a three-dimensional wave basin. Computing facilities include a LAN-based PC laboratory, together with modelling laboratories utilising Silicon Graphics workstations and high speed Pentiums, together with an extensive range of environmental modelling software.

Outstanding Local Facilities and Features
The university is located at St Lucia, a leafy suburb of Brisbane, the state capital of Queensland, the fastest growing state in Australia. Brisbane has a population of approximately one million, with a vibrant and attractive city centre, hosting an International Expo in 1988. The semi-tropical climate caters for year-round sporting and leisure activities. World famous surf beaches are within close proximity, and Moreton Bay nearby allows for the enjoyment of sailing, diving and fishing. Brisbane has a modern cultural complex of theatres, art gallery, museum, and library, with resident State theatre companies, and symphony orchestra. The city is a popular destination for touring international artists and bands.

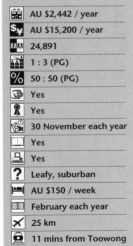

	AU $2,442 / year
	AU $15,200 / year
	24,891
	1 : 3 (PG)
	50 : 50 (PG)
	Yes
	Yes
	30 November each year
	Yes
	Yes
	Leafy, suburban
	AU $150 / week
	February each year
	25 km
	11 mins from Toowong
	6 km

Royal Melbourne Institute of Technology
Civil and Geological Engineering

Taught Courses

The Department offers Master of Engineering by research and Doctorate of Philosophy in Engineering. Courses in:
- Civil Engineering
- Environmental Engineering
- Geological Engineering

The Masters requires two years (full-time equivalent) and the PhD requires three years (full-time equivalent). Students may undertake courses on a part-time basis. There is no formal course work but students may take selected subjects in Research Method. Assessment is on thesis submission.

Research Programmes

The Department conducts applied research in five key areas shown below. Many projects arise from existing and evolving links with industry. The students supervisors may include an external industrial member.
Key areas are:
- Construction Materials (high strength concrete, composites, pavements and road materials, design and in-site performance, soils and foundations)
- Environmental Engineering (cleaner production, waste management and minimisation, hydrogeology, landfill design, site remediation and rehabilitation, sustainable development)
- Geomechanics and Geology (geo-computing, mining, geotechnical hard and soft rock engineering, petroleum exploration, slope stability remote sensing)
- Water Systems Engineering (Water & Wastewater systems analysis, maintenance, pipeline performance, hydrologic risk analysis, resources quality)
- Seismology (seismic instrumentation, earthquake analysis, hazard evaluation, donation, structures response)

Spread across these areas is research into infrastructure maintenance and project management including project risk and life cycles.

General Departmental Information

The Department is the largest in the faculty of Engineering. Post graduate students approximate 8% of student body.

The Department is located in the Central Business District with all facilities held at the one location. Students may contribute to the work of the Department in laboratory demonstration and tutoring.

Outstanding Achievements of Academic Staff to Date

The Department has 24 academic staff who have industrial and research experience across the five key areas. Students can expect comprehensive supervision which is continually being improved and would be expected to participate in seminars and produce research reports with supervisors.

Special Departmental Resources and Programmes

The Department has extensive laboratories and access to University wide facilities and support. The Seismology Research Centre is a part of the Department and conducts extensive studies in S.E. Australia and beyond. Postgraduates have access to computing facilities with full software support.

Outstanding Local Facilities and Features

Melbourne has a temperate climate and on a world survey of "most liveable" cities ranked near the top. All modern facilities, theatre, museums, libraries of other Universities are close by. Regular technical meetings are held by professional associations. Accommodation to suit all budgets is available and the city if noted for its varied restaurants reflecting the multi-cultural bodyground.

Contact
Professor Arun Kumar
Head of Department
Department of Civil and
Geological Engineering
Royal Melbourne Institute of
Technology
PO Box 2476V
Melbourne 3001
Victoria Australia

Tel   (+61) 3 9660 2181
Fax   (+61) 3 9639 0138
Email Arun.k@rmit.edu.au
WWW
http://www.civgeo.rmit.edu.au

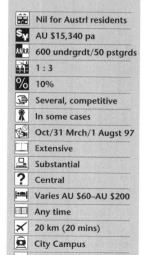

Nil for Austrl residents	
AU $15,340 pa	
600 undrgrdt/50 pstgrds	
1 : 3	
10%	
Several, competitive	
In some cases	
Oct/31 Mrch/1 Augst 97	
Extensive	
Substantial	
Central	
Varies AU $60–AU $200	
Any time	
20 km (20 mins)	
City Campus	
–	

AUSTRALIA

The University of Queensland

Contact
G J Weier
Department of Electrical and
Computer Engineering
The University of Queensland
Queensland 4072
Australia

Tel   (+61) 7 3365 3840
Fax   (+61) 7 3365 4999
Email gerri@elec.uq.edu.au
WWW
http://www.elec.uq.edu.au

Taught Courses

PhD – by research (three years full time)

Master of Engineering Science – by Research (MEngSc)

(two years full time)

Research Programmes
– Artificial Neural Networks, Intelligent Systems, Robotics
– Computer Architecture, VLSI, Digital System Design, Hardware/Software Codesign
– Microwave Systems, Radar, Antennas, Optoelectronics
– Medical Imaging, Signal Processing, Speech Processing
– Power Systems, Control Power Electronics, High Voltage Engineering
    All of the above areas of research are well-resourced and equipped and the department can offer students excellent supervision and support.
    The department offers PhD and Master of Engineering Science (by Research) projects in all of the above areas.

General Departmental Information
The department has an established reputation in neural networks research and also in digital signal processing, having recently established a Queensland node of the Co-operative Research Centre for Sensor, Signal and Information Processing headquartered in Adelaide, South Australia.
    The Hardware/Software Codesign group is the foremost in the country in this area of expertise.
    The department has strong industry links. Its advanced research program in microwave engineering led to the establishment by the Australian Federal Government of the microwave company, MITEC Australia Limited.
    The long-established High Voltage, Lightning and Insulation Laboratories enjoy a world-wide reputation in high voltage and lightning research.

Special Departmental Resources
Resources include excellent departmental computing facilities for each researcher as well as free access to the University's central supercomputers.
    Laboratory facilities include an anechoic chamber, near-field measuring facility, ground penetrating radar test pit, cytometric workstation and high voltage generator and testing systems.

Outstanding Local Facilities and Features
The University of Queensland is located in an ox-bow of the Brisbane River, about six kms from the centre of Brisbane city (population approximately 1,000,000). Brisbane is Queensland's state capital.
    Brisbane's magnificent semi-tropical climate allows for year-round sports of every type and the campus has a host of sporting and leisure facilities. Its proximity to coastal beaches which are easily accessible by electric train and bus allows enjoyment of surfing, sailing and swimming activities.
    The arts are serviced by a Cultural Centre comprising the Queensland Art Gallery, Museum and State Library, located a short walk from the city centre on the former site of Expo '88.

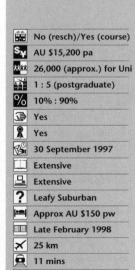

	No (resch)/Yes (course)
	AU $15,200 pa
	26,000 (approx.) for Uni
	1 : 5 (postgraduate)
	10% : 90%
	Yes
	Yes
	30 September 1997
	Extensive
	Extensive
	Leafy Suburban
	Approx AU $150 pw
	Late February 1998
	25 km
	11 mins
	6 km

The University of New South Wales
Faculty of Engineering

Taught Courses
UNSW offers over 50 courses leading to Graduate Diplomas and Master of Engineering Science in disciplines such as Biomedical; Business and Technology; Computational Science; Computer Science; Construction Management; Geotechnical; Project Management; Public Health; Structural Transport; Waste Management; Water; Communication; Electric Power; Electronics Systems and Control; Geographic and Land Information Systems; Geodesy; Photogrammetry; Aerospace; Fluid Dynamics; Heat Transfer; Computer Integrated Management; Mechatronics; Refrigeration and Air Conditioning; Design; Noise and Vibration; Environmental and Information Science.

Research Programs
The Faculty has the largest number of research students in engineering in Australia who are studying for Masters and Doctor of Philosophy Degrees. The topics include all of the above areas plus specialised programs in Biomedical Engineering; Environmental Modelling; Finite Element; Structural Analyses; Concrete Technology; Hydrology; Structural and Numerical Architecture and Graphics; Photovoltaics; Remote Sensing; Global Positioning System and Optical Fibre Technology.

General Departmental Information
The Faculty of Engineering at UNSW is the largest engineering faculty in Australia. It is also recognised as the most prominent faculty and offers the widest range of disciplines and electives including a number not available at any other university.

The staff of the engineering schools have achieved international distinction in a variety of areas, ranging from the world's most efficient photovoltaic cells to optical fibres, and co-operative research centres in Aerospace Structures; Waste Management and Pollution Control; Intelligent Manufacturing and Maritime Engineering.

Outstanding Achievements of the Academic Staff to Date
UNSW has secured more competitive grants from the Australian Research council than any other Australian institution for the past seven years. Its Faculty of Engineering has been Australia's number one research faculty for many years and again in 1995 it attracted over AUD 18 million in peer reviewed research applications.

Special Departmental Resources and Programmes
The research facilities available at UNSW are extensive and in each of the Schools of Biomedical; Civil; Electrical; Computer; Geomatic and Mechanical and Manufacturing Engineering specialist laboratories are available which contain state of the art equipment and computers.

The Faculty has strong links with industry and government through its involvement with the co-operative Research Centres which allows students access to other research facilities off campus.

Workshops are available for the fabrication of specialist equipment in all areas including Mechanical and Electronics.

Because of the wide choice the Faculty is able to offer research programs across the broad field of Engineering and Engineering Science.

Outstanding Local Facilities and Features
Located in the Eastern Suburbs of Australia most cosmopolitan city, Sydney, Capital of the State of New South Wales, UNSW is about two km from Sydney's most famous surfing beaches and is close to one of Sydney's largest public parks. Sydney is renowned internationally for its Live theatre; Opera; Sporting facilities and events and its friendly atmosphere. Sydney will host the Year 2000 Olympics!

Sydney is also famous for its harbour and there are many national parks within 100 km of the city.

Contact
Professor M S Wainwright
Faculty of Engineering
The University of New South Wales
Sydney 2052
NSW
Australia

Tel (+61) 2 9385 5000
Fax (+61) 2 9385 5456
Email
M.Wainwright@unsw.edu.au

	–
	AU $16–AU $17,000 pa
	1,500 Postgraduate
	15 : 1
	25%
	–
	Yes
	September 1997
	Yes
	–
?	City Open Architecture
	AU $100–AU $150
	I March–27 July 1997
	10 km
	20–30 mins
	7 km

AUSTRALIA

Edith Cowan University
Faculty of Science, Technology and Engineering

Contact
Anne Elam
Senior Administrative Officer
Edith Cowan University
Joondalup Campus
Joondalup Drive
Joondalup
Western Australia 6027
Australia

Tel   (+61) 9 40 05 505
Fax   (+61) 9 40 05 613
Email A.Elam@cowan.edu.au
WWW
http://www.cowan.edu.au/
ecuwis/docs/fac/ste.html.

Taught Courses
– Graduate Certificate
– Graduate Diploma of Science
– Master of Science
– Master of Engineering Science

POSTGRADUATE PROGRAMMES
The Faculty offers a range of postgraduate programmes designed in consultation with industry, government and the professions and focusing on the emerging science, technology and engineering needs of tomorrow. It provides a stimulating learning environment using the modern technology and excellent facilities.
Postgraduate Programmes include:

GRADUATE CERTIFICATE IN SECURITY MANAGEMENT

GRADUATE DIPLOMA OF SCIENCE
Archives and Records; Computer Studies; Mathematics; Security Management; Teacher Librarianship

MASTER OF SCIENCE
Computer Science; Information Science (Coursework; Research); Information Technology; Mathematics and Planning; Mathematics in Industry (Coursework); Software Engineering (Coursework)

MASTER OF ENGINEERING SCIENCE
Communication Systems; Computer Systems; Electronic Systems

PHD PROGRAMMES
Computer Science; Engineering (Communication Systems, Computer Systems, Electronic Systems); Information Science; Interdisciplinary Studies; Mathematics

General Departmental Information and Special Departmental Resources
The Faculty's School of Mathematics, Information Technology and Engineering consists of five Departments: Computer Science; Computer and Communication Engineering; Library and Information Science; Mathematics; and Security.
Also located in the School are:
– The University Research Centre for Very High Speed Microelectronic Systems, headed by internationally renowned scientist, Professor Kamran Eshraghian. The Centre specialises in hardware aspects of very fast digital processors; single-chip technology; virtual super processor; gallium arsenate integrated systems and circuits; unified GaAs/CMOS/BiCMOS technology; and virtual vision
and:
– The University Research centre, the Australian Institute of Security and Applied Technology (AISAT), which specialises in digitised access control systems; integrated campus security systems; forensic imaging, ballistics imaging, biometrics, projectile location system and smart cards. AISAT is recognised internationally as a leader in the field of security.

Outstanding Achievements of the Academic Staff to Date
– Professor Kamran Eshraghian is internationally acclaimed as the co-inventor of the bionic microchip. His pioneering work in CMOS VLSI technology has been encapsulated in standard texts now used in universities throughout the world.
– Professor Anthony Watson heads up the Computer Science Laboratory (CSL) which has international projects in Korea and France, undertaking research in Smartcard technology, computer networks, secure systems design and PC security. Professor Watson was also the designer of the Edith Cowan University Virtual Campus

Outstanding Local Facilities and Features
Edith Cowan University is both the oldest and newest higher education institution in Western Australia. Its long history of excellence goes back to the establishment of the first college at Claremont in 1902. But it is a new University, young, vigorous and innovative in its courses, its teaching and its research, and the delivery of its programmes. Currently, the University has student enrolments in excess of 20,000. Four of the University's campuses are predominantly based in the northern metropolitan area of the City of Perth, with a fifth campus based in the south west at Bunbury.

The Faculty of Science, Technology and Engineering's courses are centred on the Joondalup and Mount Lawley Campuses. Joondalup Campus is surrounded by beautiful bushland in Perth's rapidly developing northern corridor and is excellently served both by the freeway and rapid rail system. Joondalup is 20 minutes from the City of Perth and only five minutes from some of Western Australia's most attractive beaches. The Mount Lawley Campus is five minutes drive from the City of Perth, a city much acclaimed for its beautiful Kings Park, its idyllic riverside location and its pristine environment.

University of Wollongong
Faculty of Engineering

Taught Courses
The University of Wollongong is located in a delightful coastal environment on the east coast of Australia, one hours drive south from Sydney. Student enrolment in 1996 was 12,593 with 9,285 undergraduate and 3,308 postgraduates. International enrolments totalled 1,878 with students from over 70 countries.

The three departments of the Faculty are Civil and Mining; Materials and Mechanical Engineering.

UNDERGRADUATE COURSES
The Faculty has a flexible and innovative approach to engineering education, Bachelor of Engineering and Bachelor of Technology degrees are offered in the disciplines of civil environmental materials, mechanical and mining engineering.

Double degrees are offered in conjunction with the Faculties of Arts and Commerce. These undergraduate courses are accredited by the Institution of Engineers, Australia.

POSTGRADUATE COURSES
The Faculty offers one year full time coursework Master of Engineering Practice courses in:
- Bulk solids and particulate technologies
- Civil, environmental, materials and mechanical engineering
- Materials welding and joining
- Steel processing and products

The Master of Engineering Practice is designed specifically for professionals in industry and may contain modules in innovation and design, advanced techniques in project management and the latest computer techniques in engineering.

A Master of Engineering (Honours) by coursework is offered by each Department, as well as in Maintenance Management and Systems Engineering.

Research Programs
The three Departments offer both Doctor of Philosophy and Honours Master of Engineering by Research programs in a variety of research areas.

Research in the Faculty is concentrated into Institutes, Centres and Groups where a cross disciplinary teamwork approach is used:

INSTITUTES
- BHP Institute for Steel Processing and Products
- Technology and Manufacturing

CO-OPERATIVE RESEARCH CENTRES (CRC)
- CRC for Intelligent Manufacturing Systems
- CRC for Materials Welding and Joining

CENTRES AND GROUPS
- Applied Mechanics Research Group
- Biomechanical Engineering and Musculoskeletal System Group
- Centre for Bulk Solids and Particulate Technologies
- Centre for High Temperature Superconducting Materials (CSEM)
- Centre for Steel Manufacturing and Processing
- Energy Storage Materials Group
- GEM (Geo-environmental Mine) Engineering Research Centre
- Melt Processing Research Group
- Research Centre for Advanced Materials Processing (RCAMP)
- Surface Engineering Research Centre (SERC)

For more information about the activities of the Faculty of Engineering or general information about the University of Wollongong, search the WWW at: http://www.uow.edu.au/eng/eng.htm/ or telephone/fax the Faculty of Engineering office

Contact
Professor Brendon Parker
Dean of Engineering
University of Wollongong
Northfields Avenue
Wollongong 2522
Australia

Tel (+61) 42 213491
Fax (+61) 42 213143

US $2,700 pa	
US $6,250–US $7,500 pa	
12,593	
1 : 25	
15 : 85	
Yes	
–	
31 October 1997	
Yes	
Yes	
Regional University	
US $6,200 in residence	
3 March/Mid-June 1998	
80 km	
1–2 km	
90 km	

RMIT University
Department of Manufacturing Systems Engineering

Contact
MEng Course Co-ordinator
Department of Manufacturing
Systems Engineering
Bundoora East Campus
PO Box 71
Bundoora
Victoria 3083
Australia

Tel   (+61) 3 9407 6007
Fax   (+61) 3 9407 6003
Email sabu@rmit.edu.au

Taught Courses

MENG (MANUFACTURING: MANUFACTURING TECHNOLOGY STREAM)

MENG (MANUFACTURING: MANUFACTURING MANAGEMENT STREAM)

GRAD DIP (MANUFACTURING: MANUFACTURING TECHNOLOGY STREAM)

GRAD DIP (MANUFACTURING: MANUFACTURING MANAGEMENT STREAM)

Course Structure

MASTER OF ENGINEERING

There are two streams:

– **Manufacturing Technology stream**: This has a focus on the technical aspects of manufacturing such as Intelligent Materials and Processes; Computer Integrated Manufacturing; Industrial Automation and Concurrent Engineering

– **Manufacturing Management stream**: This comprises of subjects such as Manufacturing Planning and Control; Total Quality Control Systems; Manufacturing Information Systems: Workplace Organisation; Enterprise Re-engineering, and Distributed Manufacturing Systems.

Some subjects are common to both streams.

GRADUATE DIPLOMA AS AN INTERMEDIATE AWARD

For students who are enrolled in the Masters course but are unable to complete the requirements for the award of the degree of MEng, they can apply for the award of the Graduate Diploma in Manufacturing upon successful completion of 12 subjects.

General Entrance Requirements

A good first degree in Engineering is required. Applicants with other degrees or mature aged applicants with at least two years working experience in a manufacturing environment will also be considered. Subjects may be exempted on individual basis for lateral entry from other courses.

Application Deadlines

– 20th January (start of year enrolments)
– 30th June (mid-year enrolments)

Research Programmes

MEng by research and PhD degrees are also offered across a full range of manufacturing related disciplines. Current areas of research include: Robotics; Computed Integrated Manufacturing; Laser Applications; Finite Element Analysis; Manufacturing Process simulation (hardware and software); Advanced Computer Graphics; Polymer Processing and Manufacturing of Composite Materials.

General Information

RMIT University is the largest multi-level University in Australia with a history dating back to 1887. The Department is located in a parkland about 20km north of Melbourne.

The other courses offered by the Department are:

– BApp Sci in Manufacturing Operations
– BEng in Manufacturing Systems Engineering
– BEng in Manufacturing Systems Engineering/Business Administration
– BEng in Manufacturing Systems Engineering/Computer Science

MELBOURNE

Melbourne has excellent educational and recreational facilities. It is located only a couple of hours drive from the ski resorts. Beautiful mountain ranges and beaches are only minutes away from the city centre.

AU $5,000	
AU $16,780	
45	
–	
13%	
–	
–	
20 January 1997	
Very good	
Very good	
–	
AU $80–AU $110 pw	
Early March 1997	
30 km	
30 min	
20 km	

on the net **http://www.editionxii.co.uk**

Belgium

Université de Liège
ANAST

Formation
– Doctorat en Sciences Appliquées
– Ingénieur Civil des Constructions Navales
– Maîtrise en Sciences Appliquées
– Maîtrise en Gestion des Transports

Programmes de Recherche

ARCHITECTURE NAVALE
– Conception: CAO; CAD
– Hydrodynamique Navale et Offshore
– Navigation Intérieure
– Télématique Appliquée à la Navigation

ANALYSE DES TRANSPORTS
– Analyse des Systèmes de Transport Intermodaux
– Analyse Technico-Economique Entre Systèmes de Transport
– Etablissement de Plans de Transport
– Modélisation de Trafic

Informations Générales
Les étudiants sont encadrés par des spécialistes et évoluent dans un environement enrichi par de nombreux contacts avec le monde extérieur.

Réussites et Spécificités du Corps Professoral à ce Jour
Personnel à statut universitaire, chercheurs à statut définitif et provisoire, plus de 500 publications, nombreuses participations à des colloques et conférences dans le monde, membres des plus importantes organisations internationales.

Ressources et Programmes Distinctifs du Département
– Equipements informatiques performants
– Bassins à houle et d'essais de carènes
– Softwares navals et en transport
– Nombreux partenariats avec le milieu industriel et public
– Budgets publics et propres – projects CEE

Environnement Local and Régional
Toutes les facilités sont disponibles.

Contact
Professeur L J Marchal
Université de Liège
ANAST
Institut du Génie Civil
6 Quai Banning
4000 Liège
Belgique

Tel   (+32) 41 66 92 27
Fax   (+32) 41 66 91 33
Email j.marchal@ulg.ac.be

25 000 BF	
25 000 BF	
30 – 50	
–	
40%	
Non	
Non	
30 Septembre 1997	
Oui	
Oui	
–	
6 500 BF (mois)	
15 Septembre 1997	
100 km	
–	
100 km	

Katholieke Universiteit Leuven
Faculty of Engineering

Contact
Department of Metallurgy and
Materials Engineering
W. de Croylaan 2
3001 Heverloe
Belgium

Tel   (+32) 16 32 13 00
Fax   (+32) 10 32 19 00
Email
Ignace.Verpoest@mnm.kuleuve
n.ac.be
WWW
http://www.mtm.kuleuven.ac.b
e/Programs/EUPOCO/
about.htm

18 000 BEF	
18 000 BEF	
2 586	
1 : 25	
10%	
No	
No	
1 May/1 September 97	
Yes	
Yes	
Just outside the city	
4 000 BEF per week	
September/October 97	
25 km	
15 mins	
25 km	

Taught Courses
MASTER OF ARTIFICIAL INTELLIGENCE (AI)
The Artificial Intelligence programme is a multidisciplinary postgraduate programme for graduates from a broad variety of disciplines (engineering; informatics; linguistics; economy; psychology; philosophy etc). The programme, which covers one year, consists of a mandatory part (including a thesis) and at least five optional courses.
–   **Engineering and Computer Science**
This option emphasises the basic principles of AI, construction of efficient tools (programming languages and expert systems) and the application of AI techniques in such areas as robotics; speech and image processing; VLSI design and knowledge based systems.
Degree awarded: Master of Artificial Intelligence; Option: Engineering and Computer Science;
–   **Cognitive Science**
This option puts emphasis on the theoretical foundations of AI, cognitive psychology and computational representation techniques and study the principles of human interaction
Degree awarded: Master of Artificial Intelligence; Option: Cognitive Science.
    Applicants should have excellent academic credentials and be eligible for graduate study in their own country. They should hold a Master's degree or its equivalent in the field of sciences, behavioural sciences or humanities and have experience in computing concepts and practice.
    Foreign students should apply before 1 May; Belgian students before 1 September .
    All courses are taught in English.
Contact
Mrs Anne Ons
Secretariat Artificial Intelligence
Arenbergkasteel
Kardinaal Mercierlaan 94
3001 Heverlee
Belgium
Tel: (+32) 16 32 13 73   Fax: (+32) 16 32 19 82
Email: Anne.Ons@ftw.kuleuven.ac.be
WWW: http://www.cs.kuleuven.ac.be/~ml/MAI/
MASTER IN POLYMER AND COMPOSITES ENGINEERING (EUPOCO)
Since 1990 six leading European universities have been organising the European Postgraduate Education in Polymer and Composites Engineering: Katholicke Universiteit Leuven; Ecole des Mines Paris; Technische Universiteit Delft; Université Catholique de Louvain; Rheinisch-Westfalische Technische Hochschule Aachen; Imperial College London.
    The major strength of the programme is its unique combination of polymer and composites engineering. Morcover, a new modular concept has been conceived for upcoming EUPOCO year (1996 – 1997).
–   Module 1: The Basics (16 September – 27 September 1996)
–   Module 2: Advanced Polymer Science (7 October – 18 October 1996)
–   Module 3: Polymer Processing (4 November – 15 November 1996)
–   Module 4: Composites Science and Technology (25 November – 5 December 1996)
–   Module 5: Manufacturing and In Service Behaviour (6 January – 17 January 1997)
–   Module 6: Engineering with Polymers and Composites (27 January – 7 February 1997)
    Students, who follow the full programme, either as full-time student in one year, or as part-time student in two or three years and additionally carry out a research project, obtain the degree of Master in Polymer and Composites Engineering. All candidates will be evaluated on an individual student in two or three years and additionally carry out a research project, obtain the degree of Master in Polymer and Composites Engineering. All candidates will be evaluated on an individual basis before entrance.
    Occasionally students can follow a limited number of courses. These students obtain a certificate for each module they successfully complete.
    Foreign students should apply before May 1; Belgian students before September 1.
    All courses are taught in English.

Belorussia

Radioelectronics

Belarussian State University of Informatics and Radioelectronics

Taught Courses
MSc degrees are available for the following programmes:

COMPUTER DESIGN AND TECHNOLOGY

COMPUTER, SYSTEMS AND NETWORKS

ECONOMICS AND ENTERPRISES MANAGEMENT

INFORMATION PROCESSING AUTOMATED SYSTEMS

INFORMATION TECHNOLOGY SOFTWARE

MICROELECTRONICS

OPTOELECTRONIC ENGINEERING

OPERATING AND REPAIRS

RADIOELECTRONICS DEVICES DESIGN AND MANUFACTURE

RADIO EQUIPMENT TECHNOLOGY

TELECOMMUNICATION SYSTEMS

Research Programmes
PhD and DSc degrees (three years) are offered in the following areas:
- Elements; Devices and Systems Certification; Diagnostics and Testing
- Information Technologies and Control Systems
- Methodical and Socio-economical Problems of Teaching Scientific Research at High School
- Microelectronics
- New Promising Materials; Energy and Resource Saving Technologies
- Radioengineering Devices and Systems
- Simulation and Optimisation Methods of Radioelectronic Systems and Devices

General Departmental Information
The language of instruction is Russian; Besides the above programme a foundational course in the Russian language is available. The University maintains collaborate links with higher education institutions in Germany; France; Poland; China; etc.

Outstanding Local Facilities and Features
Minsk is the biggest-cultural centre of Belarus. There are nine theatres (the most popular among them is the State Academic Theatre of Opera and Ballet); 20 concert-halls (the Belarussian State Philharmonic); 14 minimum (the Belarussian Art Museum) and many other historical and cultural places to visit.

Contact
Professor Viktor M Iljin
Belarussian State University of
Informatics and Radioelectronics
6 P Brovka St
Minsk 220027
Republic of Belarus

Tel   (+7) 0172 32 0451
Fax   (+7) 0172 31 0914
Email cit@micro.rei.minsk.by

🎦	–
$¥	$3,000
👥	220
🏛	50%
%	1%
📖	–
🏃	–
🗓	1 September 1997
🗔	Yes
🖥	Yes
?	Internal
🛏	$20–$25
🗓	1 November 1997
✈	45 km
🚇	In Capital
🚋	In Capital

University of Mining and Geology
St Ivan Rilski

Taught Courses
The graduate programme of the University of Mining and Geology St Ivan Rilski offers the following degrees: Bachelor of Science (BSc) and Master of Sciences (MSc) and Doctor of Philosophy (PhD) in Advanced Materials and Mineral Synthesis; Applied Geophysics; Drilling and Oil and Gas Production; Economic Geology; Hydrogeology and Engineering Geology; Engineering Geology; Mine Automation; Mine Electrification; Mine Engineering; Mine Mechanisation; Mine Surveying and Geodesy; Mineral Processing and Underground Construction. The above courses are distributed in three of the University faculties.

Research Programmes
The postgraduate programme offers research and training in all the above specialities and the degree of Doctor of Philosophy.

General Departmental Information
The University was found in 1953 as a successor of the state Polytechnic in Sofia. It is the only higher educational institution in Bulgaria for graduate and post-graduate training of engineers for the geological exploration, mining and mineral processing industry.

Research at the University comprise both matters of general and basic scientific interest and specialised tasks in the field of Applied engineering and Technology. All the efforts are directed to Prospecting; Appraisal, Development; Exploitation and Processing of mineral raw materials and centred on solving practical technological problems as far as a large share of research is supported by industry.

Outstanding Achievements of the Academic Staff
The achievements of the academic staff are published in outstanding international magazines and journals, a number of scientific innovations are reported at international conferences and several inventions are patented at National and Foreign Patent Offices. Academic staff members are co-partners in projects under the PHARE scheme – TEMPUS, COPERNICUS, CEPUS.

Special Department Resources and Programmes
The University has got at its disposal a Museum of Paleontology; Historical Geology and Geology of Bulgaria and a Museum of Mineralogy and Petrography; a modern Computer Centre and a number of research laboratories.

Outstanding Local Facilities and Features
The University is located in a very pleasant environment at the students' campus near the centre of Sofia, the capital city of Bulgaria, where cultural life is concentrated. In the University vicinity there are sports facilities for football, tennis and winter sports.

Contact
Assoc Professor Veselin Kovachev
PhD
Vice-Rector of Training
University of Mining and Geology St Ivan Rilski
Sofia – 1100
Bulgaria

Tel (+359) 2 625 81
Fax (+359) 2 621 042
Email vvk@mgu.bg

48,000 Bulg Leva
US $3,000
2,045
0,066
0,066
3,500 Bulg Leva
–
First week of October
Yes
Yes
Single and double rooms
US $70
First week of October 97
10 km
–
15 km

BULGARIA

University of Rousse

Contact
Professor Boris Tomov
University of Rousse
8 Studentska Street
7017 Rousse
Bulgaria

Tel (+359) 82 447 268
Fax (+359) 82 451 092
     (+359) 82 455 145

Taught Courses
- Business and Management
- Design and Construction of Machines
- Electronics and Automation
- European Studies
- Electric Supply and Electro-equipment
- Industrial Design
- Information Facilities and Technologies
- Integrated Engineering
- Kinesytherapy
- Law
- Machine Engineering, Technologies and Management
- Marketing
- Mathematics and Informatics
- Mechanisation, Technology and Management in Agriculture
- Pre-School Pedagogics
- Primary School Pedagogics
- Transport Techniques and Technologies

Research Programmes
Construction and technological investigations in: Engineering; Electronics; Electroengineering; Transport; Agriculture; Production of single numbers and series of unique equipment; Organising the experimental production of highly effective parts.
The major active research projects are:
- Analysis of electromagnetic processes by some methods of vacuum plating
- Development of needs and devices for robot assembly
- Investigation of equipment for soil treatment and sowing by imported technological possibilities
- Investigation of methods and means of control of agricultural tractors

General Departmental Information
The University of Rousse was founded over 40 years ago, in October 1954, as The Higher Institute of Mechanisation and Electrical Supply for Agriculture. At present the University of Rousse develops both technical and non-technical specialists. The total number of 500 lecturers includes 150 professors and associate professors and 205 PhD specialists. There are well established international relations in the research and teaching process with leading universities in Europe. The University of Rousse is an autonomous higher school, subsidised by the State and by its own income from teaching, research, and production activities.

Outstanding Achievements of the Academic Staff to Date
Seven lecturers-professors are members of the Academy of Science of New York.

Special University Resources and Programmes
All premises, facilities, laboratory equipment, students' hostel and canteen are provided for all students.

Lv 19,000 per semester
US $1,250–US $2,000
6,000
1 : 6
–
–
For home students only
1 September 1997
Available
Available
? –
–
15 September 1997
380 km
5 hours
380 km

Sofia Technical University

Taught Courses
Automatics and Systems Engineering; Electrical Power Production and Electrical Equipment; Electronics; Electrical Engineering; Industrial Management; Computer Systems. Communications and Communications Technologies; Machine Building Engineering and Manufacturing Equipment; Mechanical and Device Engineering; Textile Engineering; Thermal and Nuclear Power Engineering; Heat Transfer Engineering and hydraulic Machines; Transport Machinery and Transportation Equipment; Mechanical Engineering (in German); Electrical Engineering (in French); Industrial Engineering (in English).

Research Programmes
R&D production activities at TU are vital for the development of the educational process at TU. They are organised in the following division: R&D Sector; Small - batch Production Units and the Pilot Plan.

The various types of activity are covered by the following contracts: Fundamental research projects; State – financed through the Research and Development Fund; International scientific projects; Joint activities with external institutions; Technical evaluation; Application of research products in industry; Provide experimental implementation and manufacture within TU research Projects; Production facilities in the area of machine building; Electronic equipment; Laser technologies and NC machines.

General Department Information
The TU of Sofia in an autonomous institution.

The managing bodies of the TU are as follows:
- **The General Assembly**: It composed of all representatives of the Assistant; Professors; Full time PhD students; Students representatives and administrative officers of the TU.
- **The Academic Council**: Sets guidelines for the Educational; Research and Personnel management policy of the TU.
- **The Rector**

The main division for organisation of institution and research is the Faculty. The University comprises 10 main faculties. The primary unit in the TU educational and research activities is the Department. Its managing bodies are the Department Council and the Head of Department. TU has branches in the towns of Plovdiv and Sliven as well as several affiliated institutions.

Outstanding Achievements of the Academic Staff to Date
The TU is founder of the Association for business Co-operation between the international relations departments of the Higher Technical schools from Central and Eastern Europe. A specific form of co-operation between academic from the TU and renowned universities in Eastern and Western Europe are join projects under COST; TEMPUS; COPERNICUS; ERASMUS programmes. The TEMPUS projects, focusing on curriculum design to acquire the European standards of education, proved especially fruitful. In the period 1992 – 96 their total number came up to 28, the funds received amounted to ECU over 1.5 million.

Contact
Professor Dr Venelin Jivkov
Technical University of Sofia
1000 Sofia
Darvenitsa
Bulgaria

Tel   (+359) 2 62 30 73
Fax   (+359) 2 68 53 43

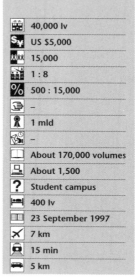

	40,000 lv
	US $5,000
	15,000
	1 : 8
	500 : 15,000
	–
	1 mld
	–
	About 170,000 volumes
	About 1,500
	Student campus
	400 lv
	23 September 1997
	7 km
	15 min
	5 km

Czech Republic

General Engineering

Czech Technical University in Prague
Engineering

Taught Courses
Program taught in English leading to PhD degree are offered at the Faculties of the Czech Technical University in Prague:
– Architecture
– Civil Engineering
– Electrical Engineering
– Masaryk Institute of Advanced Studies and Klokner Institute
– Mechanical Engineering
– Nuclear Sciences and Physical Engineering
– Transportation Science

Research Programmes
Research is undertaken in all the basic disciplines taught at the university and in many interdisciplinary areas. The major part of our research can be characterised as mission orientated and applied research. Research is carried in all six faculties and two institutes.

General Departmental Information
The Czech Technical University in Prague with its 2,500 members of academic staff working in about 150 research groups can be considered as one of the largest research institutions in the Czech Republic. Generally, each research team belongs to a department and its activities are financed from grants for projects or from the faculty budget through the department.

Outstanding Achievements of the Academic Staff to Date
– Prof. Ing. J Valenta, DrSc: Research results in the field of Biomechanics of a Man
– Prof. RNDr K.Rektorys, DrSc: Survey of Applicable Mathematics
– Prof. Ing. H Havlícek, DrSc: Hilbert Space Operators in Quantum Physics

Special Departmental Resources and Programmes
Main additional resources are based on European Union programmes like TEMPUS, COPERNICUS, COST, further grants are awarded by UNESCO, NATO, NSF and are gained through co-operation with big industrial companies.

Outstanding Local Facilities and Features
The Old Town – the National Theatre, the Old Town Hall, Gothic T"n Church, the Bethlehem Chapel, the Charles Bridge, the New Town – the Venceslas Square, the National Museum, the Lesser Town – the Prague Castle, Neruda s Street...

Contact
Doc. Ing. B Rîha, CSc.
Czech Technical University
in Prague
Zikova 4
166 35 Prague 6
Czech Republic

Tel (+42) 2 2435 3462
Fax (+42) 2 2431 1042
Email riha@vc.cvut.cz

US $2,500/per semester
US $2,500/per semester
600 (ft)/ 550 (pt)
1 : 10 (whole university)
–
86
498
March or September 97
Yes
Yes
Central and Satellite
US $15
September 97
10 km
–
–

Finland

Ecole Centrale de Nantes / Université de Nantes

Formation Doctorale Automatique & Informatique Appliquée

Formation Enseignée

DEA AUTOMATIQUE & INFORMATIQUE APPLIQUÉE

Ce diplôme de 3ème cycle forme les étudiants aux techniques avancées et aux résultats de recherche dans le domaine général de la conduite de systèmes industriels complexes. Le DEA AIA dispense aux étudiants une formation spécialisée dans les domaines de l'automatique fondamentale; du traitement du signal; de la robotique; des systèmes à événements discrets; et de l'informatique embarquée temps réel.

Un des objectifs du DEA est de préparer les étudiants à intégrer une équipe de recherche universitaire pour y préparer une thèse de doctorat. Les étudiants qui soutiennent une thèse dans le cadre de cette formation peuvent ensuite postuler sur un emploi de Maître de Conférences dans les disciplines de l'automatique; du traitement du signal; de l'informatique; de la robotique ou de la productique.

Un second objectif du DEA AIA est de permettre aux étudiants l'accès aux départements "Etudes et Recherche" d'entreprises innovantes pour y participer à des études avancées. Les entreprises concernées sont celles de production en continu de matière; d'énergie; ou les services de conception d'équipements de haute technologie; ou les sociétés de service à haut niveau d'expertise dans les domaines de l'automatique; de la productique; de l'informatique validées et sûres de fonctionnement.

Organisation du DEA

Quatre filières de formation théorique sont proposées aux étudiants:
- Commande des systèmes
- Informatique Appliquée
- Robotique
- Traitement du Signal

Le volume horaire total de cours magistral suivi dans chaque filière est de 130 heures, dispensées entre la mi-Novembre et le mois de Mai.

La partie stage de recherche du DEA s'effectue généralement dans un des deux laboratoires support de cette formation : le LAN et le LISA. Mais elle peut également s'effectuer dans un autre laboratoire, ou à l'étranger. Dans tous les cas, elle est dirigée par une personnalité scientifique de la Formation Doctorale.

Programme de Recherche de la Formation Doctorale

Après le DEA, une thèse (PhD) peut être préparée dans la Formation Doctorale AIA.

Tout doctorant bénéficie d'un support financier qui peut avoir plusieurs origines (Ministères; CNRS; Entreprises; Région; etc).

Les thèmes de recherche développés en thèse sont les suivants :
Structure des systèmes linéaires; Systèmes à retards; Structure et commande des systèmes non linéaires; Observateurs en non-linéaire; Systèmes à paramètres répartis; Traitement du signal; Détection de ruptures; Systèmes non-stationnaires; Statistiques d'ordre supérieur; Systèmes temps réel; Langages réactifs asynchrones; Systèmes de transition; Modèles et preuves de logiciels embarqués; Ordonnancement; Configuration, placement; Protocoles de communication; Analyse dynamique des systèmes à événements discrets; Algèbre (max,+); Robotique; Modélisation de robots; Identification et commande de robots; Robotique mobile.

Ces travaux s'appuient sur de nombreux contrats de recherche et industriels, nationaux et européens.

Corps Professoral et Unités de Recherche

L'enseignement magistral est dispensé par les enseignants de l'Ecole Centrale de Nantes; de l'Université de Nantes; de l'Université d'Angers; de l'Ecole des Mines de Nantes; et par des chercheurs du CNRS. En outre, de nombreuses personnalités présentent leurs plus récents résultats scientifiques dans le cadre des séminaires organisés par le LAN et le LISA. La recherche s'appuie sur :
- le Laboratoire d'Automatique de Nantes (LAN), unité de recherche du CNRS depuis 1968, et qui regroupe environ 50 enseignants-chercheurs, sept chercheurs CNRS et 50 doctorants;
- le Laboratoire d'Ingénierie des Systèmes Automatisés d'Angers (LISA), unité de recherche reconnue par le Ministère de l'Enseignement Supérieur et de la Recherche, et qui regroupe environ dix enseignants-chercheurs et dix doctorants.

Contact
Jean-Pierre Elloy
Ecole Centrale de Nantes
BP 92101
44 321 Nantes cedex 03
France

Email Jean-Pierre.Elloy@ lan.ec-nantes.fr
WWW
http://www.ec-nantes.fr/

Secrétariat
Geneviève Neveu

Tél (+33) 2 40 37 25 11
Fax (+33) 2 40 37 25 22
Email Genevieve.Neveu@lan.ec-nantes.fr

🏨	2 000 FF
$¥	2 000 FF
👥	40 (DEA) + 35 (PhD)
📊	60%
%	5 – 10%
	Oui
	Oui
	1 Juillet 1997
	Oui
	Oui
?	Urbain
	500 FF
	15 Novembre 1997
✈	20 km
	2 heures (TGV)
🚌	400 km

Ecole Universitaire d'Ingénieurs de Lille

Contact
M Christian Doremus
Mme Guislaine Petit
M Jian-Fu Shao
EUDIL
Université des Sciences et
Technologies de Lille
59 655 Villeneuve d'Ascq cedex
France

Tel   (+33) 3 20 43 46 08
Fax   (+33) 3 20 43 43 35
Email Scola@eudil.univ.-lille1.fr

Formation

DIPLÔME D'INGENIEUR

Trois ans en formation initiale et formation continue – cinq specialités:
– Géotechnique – Génie Civil (travaux publics, ouvrages souterrains, reconnaissance des sols)
– Informatique - Mesures – Automatique (micro-ondes, éléctronique de puissance, logiciels, automatismes)
– Instrumentation (formation technico-commerciale)
– Mécanique (conception, modélisation de systèmes)
– Science des matériaux (métaux, polymères, composants électroniques)
Admission
– Bac + 2
– Concours écirts + entretien pour les Classes Préparatoires aux Grandes Ecoles
– Inscription: Décembre 1996 – Janvier 1997
– Dossier + entretien pour les diplômes universitaires scientifiques (DEUG, DUT, titres étrangers admis en équivalence) Inscirption: 1Mars 1997 – 15 Mai 1997
Partenariat
Partenaire du Réseau Eiffel avec le CUST (Institut des Sciences de l'Ingénieur de Clermont-Ferrand) et l'ISIM (Institut des Sciences de l'Ingénieur de Montpellier)

Programmes de Recherche

DEA (DIPLÔME D'ETUDES APPROFONDIES) ET FORMATIONS DOCTORALES EN GÉNIE CIVIL

Rhéologie des géomatériaux et stabilité des ouvrages souterrains: Matériaux, habitat et génie urbain; Valorisation des déchets et sous-produits industriels
Admission
– Maîtrise de Génie Civil; Mécanique et Physique; Diplôme d'Ingénieur; Diplômes étrangers admis en équivalence
– Inscription: avant le 1er Juillet 1997
Partenariat
Université des Sciences et Technologies de Lille; Université d'Artois; Ecole Centrale de Lille

INSTRUMENTATION ET ANALYSE AVANCÉES

Instrumentation scientifique dans le domaine de la physico-chimie, biomédicale et agro-alimentaire
Admission
Maîtrises scientifique, Diplôme d'Ingenieur; Diplômes étrangers admis en équivalence. Inscription: avant le 1er Juillet 1997
Partenariat
Nombreux laboratoires de l'Université des Sciences et Technologies de Lille; Université du Littoral; Institut d'Electronique et Microélectronique du Nord; Ecole Nationale Supérieure des Arts et Industries Textiles.

MASTÈRE SPÉCIALISÉ EN GÉNIE E L'EAU ACCRÉDITÉ PAR LA CONFÉRENCE DES GRANDES ECOLES

Admission
Diplôme d'Ingénieurs à caractère scientifique ou diplômes étrangers admis en équivalence.
Inscription avant le 01 Septembre 1997.

Université de Valenciennes et du Hainaut Cambrésis
Département d'Electronique

Formation
DEA Electronique Ultrasons et Imagerie
Formation des images, transformée de Fresnel, transformée de Fourier – Traitement de signal et de l'image – Propagation des ondes dans les milieux anisotropes.

Option Ultrasons: théorie et applications au sonar, au domaine médical et au contrôle industriel.

Option Imagerie: théorie et applications des systèmes numériques et analogiques d'acquisition, de transmission et de restitution des images.

Programmes de Recherche
– Traitement de signal; Processeurs acousto-optiques; Interactions électro-acousto-optiques dans les supers réseaux; Vidéoprojection; Transmission et compression de signaux en vidéo numérique.
– Evaluation non destructive par ultrasons et applications au GBM; Interactions ultrasons-matière; Sonochimie; Acoustique sous-marine; Moteurs ultrasonores; Soudure par ultrasons; Céramiques pour l'acoustique et l'optique.

Informations Générales
Les recherches sont proposées dans différents laboratoires, en particulier dans les départements du nouvel Institut d'Electronique et de Micro-électronique du Nord (DOAE, ISEN et DHS).

Renseignements Utiles à l'Inscription
– Sélection sur dossier des candidats justifiant d'un diplôme scientifique de niveau au moins égal à BAC+4
– Dépôt des dossiers avant le 30 Juin 1997

Formation
DESS Mesure et Traitement du Signal
S'appuyant sur l'activité du laboratoire OAE (Opto–Acousto–Electronique) intégré à l'Institut d'Electronique et de Microélectronique du Nord (IEMN), ce DESS propose des enseignements professionnels et appliqués à la mise en œuvre de composants nouveaux tels que les "Digital Signal Processors" et leur application au domaine de l'instumentation et de la mesure.
La formation se veut pluridisciplinaire et concerne les activités industielles suivantes:
– Développement et fabrication d'instrumentation électronique
– Introduction de moyens technologiques nouveaux dans la mesure (contrôle non destructif, radar, sonar)

Admission
– Sélection sur dossier à retirer au CIO de L'Université de Valenciennes
– Cette formation s'adresse aux titulaires d'une maîtrise dans les domaines de l'électronique, du traitement de signal, et des télécommunications

Stage
Un stage d'une durée minimale de cinq mois est effectué dans une entreprise travaillant dans le domaine de la mesure et du traitement de signal.
Des possibilités de stage dans des grandes entreprises allemandes sont proposées.

Contact
DEA Electronique Ultrasons et Imagerie
Professeur C Bruneel
ISTV
Université de Valenciennes
BP 311
59 304 Valenciennes Cedex
France

Tél   (+33) 3 27 14 12 38
Fax   (+33) 3 27 14 11 89

DESS Mesure et Traitement du Signal
B Nongaillard
Laboratoire OAE
F 59 326 Valenciennes Cedex

Tél   (+33) 3 27 14 12 40
Fax   (+33) 3 27 14 11 89

🏫	– / 1 600 FF
💲	– / 1 600 FF
👥	– / 11 000
🏢	– / –
%	– / 25%
🎓	– / Oui
🎗	– / Oui
📅	– / 11 September 1997
💻	Oui
🖥	Oui
?	Urbain
🛏	– / 450FF–600 FF
📅	– / 4 October 1997
✈	40 km
🚉	2 heures
🚌	200 km

Université Blaise Pascal – Clermont-Ferrand
Ecole Doctorale Sciences pour l'Ingénieur

Contact
Secrétariat de l'Ecole Doctorale
UFR de Recherche Scientifique
et Technique
24 avenue des Landais
63 177 Aubière Cedex
France

Tél   (+33) 4 73 40 72 64
Fax   (+33) 4 73 40 72 62

Taught Courses
The Blaise Pascal University of Clermont-Ferrand provides both theoretical and practical training in the principal areas of Engineering Sciences in association with CUST – Institut des Sciences de l'Ingénieur; IFMA – Institut Français de Mécanique Avancée; and Auvergne University.

L'Ecole Doctorale Sciences pour l'Ingénieur results from the grouping of three DEA (Diplômes d'Etudes Approfondies – one year) and Doctorates (PhD – three years) providing advanced studies in co-operation with numerous laboratories:

ELECTRONICS AND SYSTEMS
Three options: Vision for Robotics; Materials and Devices for Electronics; Electromagnetism.

COMPUTER SCIENCE
Three options: Computer Science and Industrial Systems; Database Management Systems; Image Analysis and Medical Applications.

MATERIALS – STRUCTURE – RELIABILITY
Mechanical Engineering; Civil Engineering.

Research Programmes
Students perform the PhD in one of the following laboratories:
– Laboratoire des Sciences et Matériaux pour l'Electronique et d'Automatique (LASMEA) – URA CNRS 1793
– Arc Electrique et Plasma Thermiques – URA 828
– Electrotechnique de Montluçon – EA 989
– Laboratoire de Recherche et d'Applications en Mécanique Avancée (LARAMA) – JE 224
– Génie Civil – EA 990
– Informatique – EA 991

The research themes developed at Blaise Pascal University in the field of Engineering Science cover large active research areas in the above mentioned programmes. The Clermont-Ferrand scientific community is involved in a large number of European research projects. The Erasmus and Templus exchanges are numerous.

General Information
During the year of DEA, the teaching consists of basic core programmes with additional subjects with respect to the different options.

Common courses are provided in the field of innovation in the firms, the knowledge of industrial activities, the problem of industrial quality, the intellectual and industrial properties.

	–
	–
	200
	–
	–
	Yes
	Yes
	June–July 1997
	Yes
	–
?	Urban
	–
	October 1997
	–
	–
	–

Université de Technologie de Compiègne

Diplômes d'Etudes Supérieures Spécialisées

Actuellement l'Université de Technologie de Compiègne propose cinq DESS.

Les Diplômes d'Etudes Supérieures Spécialisées – DESS – sanctionnent une formation préparant directement à la vie professionnelle. Ils peuvent donc être préparés en formation initiale ou continue.

Les enseignements s'articulent autour de cours de niveau 3ème cycle et d'un stage professionnel. Ce dernier donne lieu à la rédaction d'un mémoire et à une soutenance orale présentée devant un jury.

Technologies Biomédicales Hospitalières

La technicité croissante des hôpitaux rend indispensable la présence de spécialistes susceptibles d'assurer un dialogue autant avec le monde médical et soignant qu'avec l'administration gestionnaire des équipements de santé et les entreprises biomédicales.

Programme: Physiologie; Sécurité électrique et sûreté des systèmes complexes; Imagerie médicale (radiologie conventionnelle et numérique, échographie, IRM, médecine nucléaire); Laboratoires d'analyses et explorations fonctionnelles (principes, équipements, médecine nucléaire); Traitements et soins (bloc opératoire, réanimation, monitoring, radiothérapie, lasers, hémodialyse); Management hospitalier (informatique et maintenance, gestion des hôpitaux, hygiène hospitalière); Initiation à la gestion des entreprises; Culture générale et langue anglaise.

Design Industriel – Conception de Produits

Les diplômés s'intègrent dans des équipes de conception industrielle, bureaux d'études, recherche et développement ou gestion de l'innovation dans les entreprises.

L'UTC a créé autour de la conception des produits, de l'ergonomie et du design industriel, un pôle de développement technologique et de recherche en liaison avec l'industrie.

Programme: Conception de produits; Identité visuelle de l'entreprise; Composantes esthétiques et plastiques de produits industriels; Méthodologie et analyse de la valeur; Evolution des matériaux en conception de produits; Architecture des produits et management de la conception; Ergonomie du produit; Analyse des produits de consommation; Outils informatiques; Etudes de cas industriels sur cahier des charges conceptuel.

Informatique pour la Ville

L'objectif de cette formation est de former des diplômés ayant une vision très large des problèmes informatiques des collectivités locales. Formation soutenue financièrement et pédagogiquement par le groupe Lyonnaise des Eaux, la société IBM et la société DIGITAL.

Programme: Système d'information: Bases de données, Réseau informatique, Nouvelles technologies (traitement d'images et télédétection, interfaces homme-machine); Systèmes d'informations géographiques: mise en œuvre d'un SIG; Connaissance des collectivités locales; Gestion de projet; Anglais.

Gestion de la Technologie et de l'Innovation

Le DESS GTI répond à la demande des industriels en cadres capables de conduire le lancement de nouveaux produits ou la mise en place de nouveaux procédés: Ingénieurs d'études; Chefs de produits; Chefs de projets; Chargés d'études ou consultants en innovation technologique; Ingénieurs commerciaux; Ingénieurs d'affaires.

Programme: Economie de l'innovation et politiques publiques, stratégies d'entreprise, sciences, techniques et société; Conduite de l'innovation dans l'entreprise, gestion de projet, gestion financière de l'innovation; Conception de produits et qualité industrielle; Marketing high tech; Valorisation de la recherce et transferts; Création et développement de PME innovantes.

L'Association Nationale de la Recherche Technique (ANRT) et Schneider Electric participent directement au fonctionnement du DESS.

Physicochimie des Surfaces, Systèmes Colloïdaux et Fluides Composites

La diversification et la croissance du marché mondial des spécialités chimiques (cosmétiques, vernis, colles et adhésifs, textiles, fluides d'usinages et de traitement de surfaces, etc) imposent aux entreprises d'améliorer les performances des formulations et d'en concevoir de nouvelles.

Programme: Chimie des agents tensio-actifs et comportement en solution; Eléments de chimie macromoléculaire et physicochimie des polymères en solutions; Thermodynamique des systèmes colloïdaux et des changements de phases; Physicochimie des systèmes dispersés; Rhéologie des systèmes colloïdaux; Aspects physicochimiques de la mise en oeuvre et des propriétés d'usage (mouillage, étalement, filmification, adhésion, coalescence, lubrification, flottation); Technologie de mise en oeuvre et d'application, traitement des informations et modélisation, toxicologie, écotoxicologie, réglementation.

Contact
UTC
Centre de Recherche
de Royallieu
BP 529
60 205 Compiègne Cedex
France
Fax   (+33) 3 44 20 48 13

Technologies Biomédicales Hospitalières
Georges Chevallier
Tél   (+33) 3 44 23 43 85

Design Industriel – Conception de Produits
Danielle Quarante
Tél   (+33) 3 44 23 45 59

Informatique pour la Ville
Gérard Govaert
Tél   (+33) 3 44 23 44 86

Gestion de la Technologie et de l'Innovation
François Romon
Tél   (+33) 3 44 23 46 13

Physicochimie des Surfaces, Systèmes Colloïdaux et Fluides Composites
Danièle Clausse
Tél   (+33) 3 44 23 46 14

🗓	3 000 FF
S¥	3 000FF
👥	15–20
🏛	20%
%	10%
✍	–
👤	–
🏠	Fin Juin 1997
🖥	Oui
🖥	Oui
?	–
🛏	800 FF–2 000 FF pm
📅	15 Sept–15 Oct 1997
✈	50 km
🚉	1 heure
🚐	80 km

Université de Franche-Comté
Ecole Doctorale Sciences pour l'Ingénieur et Microtechniques

Contact
Professeur G Lallement
Responsable de l'Ecole Doctorale
Ecole Doctorale SPIM
Université de Franche-Comté
Laboratoire de Mécanique
Appliquée
24 chemin de l'Epitaphe
25 030 Besançon cedex
France

Tél   (+33) 3 81 40 29 06
Fax   (+33) 3 81 40 29 01

🏛	1 000 FF
💲¥	1 000 FF
👥	240
🎓	0.3
%	10%
💳	Oui
🧍	Oui
📅	1 Juillet–1 Sept 97
⬜	–
🖥	–
?	Urbain
🛏	200 FF
📅	1ère semaine d'Octobre
✈	150 km
🚌	3 heures
🚆	400 km

DEA ACOUSTO-**O**PTO-**E**LECTRONIQUE ET **M**ÉCANIQUE DES **S**TRUCTURES
Contact:
> Professeur J P Goedgebuer
> Laboratoire d'Optique, UFRST
> 25030 Besançon – France
> Tél: (+33) 3 81 66 64 01

DEA ANALYSES ET **O**PTIMISATION EN **E**NERGÉTIQUE
Contact:
> Professeur J P Prenel
> Institut de Génie Energétique
> Avenue Jean Moulin
> 90000 Belfort – France
> Tél: (+33) 3 84 57 82 00

DEA SCIENCES DES **M**ATÉRIAUX, **M**ÉCANIQUE ET **M**ÉCANIQUE DES **S**URFACES
Contact:
> Professeur C Lexcellent
> Laboratoire de Mécanique Appliquée R Chaléat – UFRST
> 24 Chemin de l'Epitaphe
> 25030 Besançon – France
> Tél: (+33) 3 81 40 29 03

DEA INFORMATIQUE, **A**UTOMATIQUE ET **P**RODUCTIQUE
Contact:
> Professeur A Bourjault
> Laboratoire d'Automatique
> 26 rue de l'Epitaphe
> 25030 Besançon Cedex – France
> Tél: (+33) 3 81 40 28 02

PROCÉDÉS ET **T**RAITEMENT DE L'**E**NERGIE **E**LECTRIQUE
Contact:
> Professeur J M Kaufmann
> IUT Belfort
> Rue Engel Gros
> 90016 Belfort – France
> Tél: (+33) 3 84 58 77 00

Programmes de Recherche
L'Ecole doctorale SPIM regroupe cinq DEA s'appuyant sur 23 laboratoires et équipes d'acceuil et sur environ 200 chercheurs.

Ces laboratoires encadrent les stages de DEA et les Thèses de Doctorat en Sciences pour l'Ingénieur et Microtechniques dans les domaines de pointes multidisciplinaires: Acoustique; Automatique; Electronique; Energétique; Informatique; Mécanique des Matériaux; Mécanique des Structures; Mise en forme des Matériaux; Productique; Optique et Opto-Electronique; Robotique; Thermique; Technologies; Microélectroniques et Microtechniques.

Informations Générales
– Conditions d'admission dans les DEA: Maîtrise Scientifique; Diplôme d'Ingénieur; Titre français ou étranger admis en équivalence
– Sélection des candidats sur dossier (première session: 1 Juillet 1997; deuxième session: 1 Septembre 1997)
– Pour chaque DEA: enseignements théoriques de 125 à 150 heures et stage de 600 à 800 heures dans l'un des laboratoires d'accueil
– Effectif des DEA: de 25 à 70 étudiants

Ressources et Programmes Distinctifs du Département
Specialités: Secteur de Recherche des Sciences pour l'Ingénieur et Microtechniques.
Partenatiats: Ecole Nationale Supérieure de Mécanique et des Microtechniques; Institut des Microtechniques de Franche-Comté; Réseau National en Microtechologies.

on the net **http://www.editionxii.co.uk**

Institut National des Sciences Appliquées de Lyon

Taught Courses

DEA GÉNIE DES MATÉRIAUX

DEA Microstucture et Comportement Mécanique et Macroscopique des Matériaux:
Génie des Matériaux – One year
Each student follows:
– Common compulsory courses (about 100 h)
– Optional courses in three areas (60 h)
   - •Structual metallurgy
   - •Complex material systems
   - •Mechanical and macroscopic behaviour of materials
– Laboratory research individual work

Research Programmes

THESIS – THREE YEARS

After the DEA (or equivalent level) a thesis can be prepared in one of the laboratories attached to the DEA at INSA (three laboratories) or in other authorised institutions: Ecole Centrale de Lyon (three laboratories) and Université Lyon I (three Laboratories).

General Departmental Information

LEVEL REQUIRED FOR DEA

Engineer or equivalent to french "Maîtrise" (four or five years of graduate studies) in the fields of Materials; Physics; Chemistry or Mechanics.

LEVEL REQUIRED FOR THESIS

DEA or Masters

Outstanding Achievements of Academic Staff to Date

– Numerous publications in International Reviews and International Congresses on Materials (Metals and Alloys, Ceramics, Polymers, Composites)
– Organisation of National and International Congresses: Journées Nationales du Groupe Français de la Céramique, Lyon, February 96; European Symposium on Martensitic Transformation and Shape Memory Properties, Aussois, September 1991; Third International Conference on Intelligent Materials, Lyon, June 1996.

Special Departmental Resources and Programmes

Extensive equipments for elaboration, testing and investigation of materials (metals, ceramics, composites, polymers).

Outstanding Local Facilities

Usual facilities of a one million inhabitants town (museums, theatres etc).

Contact

Gérard Guenin
DEA Génie des Matériaux
Bâtiment 502
INSA
69 621 Villeurbanne
France

Tel   (+33) 4 72 43 82 45
        (+33) 4 72 43 83 85
        (secretariat)
Fax   (+33) 4 72 43 88 30
Email bernavon@insa.insa-lyon.fr

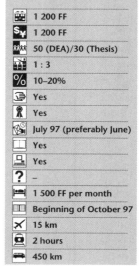

1 200 FF
1 200 FF
50 (DEA)/30 (Thesis)
1 : 3
10–20%
Yes
Yes
July 97 (preferably June)
Yes
Yes
–
1 500 FF per month
Beginning of October 97
15 km
2 hours
450 km

Université de Metz – ENIM
ISGMP

Contact
Marcel Berveiller
Université de Metz – ISGMP
Diplôme d'Etudes Approfondies
Matériaux – Mécanique –
Structures – Procédés
LPMM – ISGMP
Ile du Saulcy
57 045 Metz cedex 1
France

Tél   (+33) 3 87 31 52 60
Fax   (+33) 3 87 31 53 66

Formation

DEA MATÉRIAUX – MÉCANIQUE – STRUCTURE – PROCÉDÉ

Le DEA Matériaux – Mécanique – Structure – Procédé vise à donner une solide formation de base sur les matériaux de structure, et le comportement thermomécanique des matériaux. Il vise également à compléter ces enseignements de base par une formation au calcul des structures et aux procédés de fabrication mécanique. Son contenu a été conçu pour conforter et élargir les connaissances acquises par des étudiants ayant suivi des formations à dominante Mécanique ou Matériaux.

Programmes de Recherche

La recherche s'appuie sur:
– l'ensemble des laboratoires regroupés au sein de l'Institut Supérieur de Génie Mécanique et Productique, structure de recherche commune à l'Université de Metz et à l'ENIM
– les laboratoires de Reims (Mécanique et Matériaux)

Réussites et Spécificités du Corps Professoral à ce Jour

DÉBOUCHÉS

Les débouchés industriels se situent dans le vaste domaine Mécanique–Matériaux, allant de la conception de produits à leur fabrication (Production) en passant par l'optimisation et le choix des matériaux.

🏫	–
$¥	–
👥	32
🏛	–
%	30%
📖	Oui
🎓	Oui
🕐	Fin juin 1997
🖥	–
💻	–
?	–
🛏	–
🗓	Mi-Septembre 1997
✈	15 km
🚆	350 km
🚌	–

Université Henri Poincaré – Nancy I
Ecole Nationale Supérieure des Technologies et Industries du Bois

Directeur de l'Ecole
Professeur Xavier Deglise

Formations
- DESS Matériaux Bois et Mise en Oeuvre dans la Construction
- DEA Sciences du Bois en cohabilitation avec l'Université de Bordeaux I et l'Ecole Nationale du Génie Rural et des Eaux et Forêts

DESS
Objectif: Faire construire avec du Bois.

Modules d'Enseignements
400 heures de cours complétés par un projet et un stage en entreprise.
Xylotechnologie; Architecture bois et spécificités; Résistance des matériaux; Procédés de transformation et de mise en oeuvre du bois; Conception bois; Outils informatiques dédiés; Matériaux bois dans la construction.

Informations Générales
Formation bénéficiant du soutien du CNDB (Comité National pour le Développement du Bois).

Renseignements Utiles
Formation réservée en priorité aux architectes et aux étudiants en Génie Civil. Recrutement sur dossier et entretien.

DEA
Objectif: Connaissances approfondies sur le matériau bois et sa mise en oeuvre

Modules d'Enseignement

TRONC COMMUN
Xylologie et variabilité; Physique et mécanique du bois; Chimie du bois.

OPTIONS
Rhéologie du bois; Mécanique des composites à base de bois; Rupture et endommagement; Capteurs et optimisation des débits; Mécanique de la coupe; Chimie et biologie de la préservation; Transferts simultanés de masse d'énergie appliqués au séchage; Adhésion, collage et traitement de surface; Valorisation chimique et énergétique du bois; Modélisation de la qualité du bois en fonction de la croissance; Économie de la filière bois et des entreprises.

Programmes de Recherche
Dix laboratoires développant des recherches sur huit thèmes: Qualité des bois et sylviculture; Mécanique et structures; Séchage; Productique; Préservation; Valorisation des co-produits; Collage; Finitions; Economie.

Informations Générales
DEA national associant la plupart des équipes universitaires et publiques développant des recherches sur le Matériau Bois.

Renseignements Utiles
- Inscriptions à partir du mois d'Avril sur dossier et entretien
- Date limite de dépôt de dossier de prise de contact: 30 Juin 1997
- Date limite d'inscription administrative: 30 Septembre 1997

DESS Contact
Pascal Triboulot
Gilles Duchanois
ENSTIB
27 Rue du Merle Blanc
BP 1041
88 051 Epinal cedex 9
France

Tél (+33) 3 29 81 11 50
Fax (+33) 3 29 34 09 76

DEA Contact
Pr André Zoulalian
ENSTIB Université Henry
Poincaré, Nancy I
BP 239
54 506 Vandœuvre Les Nancy cedex
France

Tél (+33) 3 83 91 20 57
Fax (+33) 3 83 91 21 02

📷	–
$¥	–
👥	20 (DESS)/40 (DEA)
🏫	–
%	–
📖	–
👤	–
🕐	30 Juin 1997
📇	–
💻	–
?	–
🛏	–
📅	30 Septembre 1997
✈	–
🏧	–
🚌	–

Institut National Polytechnique de Lorraine (INPL)

Contact

Institut National Polytechnique de Lorraine (INPL)
2 Avenue de la Forêt de Haye - BP 3
54 501 Vandoeuvre cedex
France

Tél   (+33) 3 83 59 59 59
Fax   (+33) 3 83 59 59 55

Liste des DEA

DEA Automatique et Traitement Numérique du Signal (ATNS)
Responsable: R Husson
Cohabilitation: UHP; Nancy I et Université de Reims
Objectifs: Formation des Cadres Scientifiques pour et par la recherche dans un des pôles principaux de recherche en France; dans le domaine de l'automatique; du traitement numérique du signal et du diagnostic.

DEA Biotechnologies et Industrie Alimentaires
Responsable: J M Engassser
Cohabilitation: –
Objectifs: Formation théorique et pratique de haut niveau pour les industries biologiques et agro-alimentaires satisfaisant aux demandes actuelles et prévisibles du secteur aval.

DEA Sciences Agronomiques
Responsable: A Guckert
Cohabilitation: Université de Metz
Objectifs: Formation de haut niveau par la recherche en sciences agronomiques visant une insertion dans les secteurs de l'agriculture et des industries connexes; de l'aménagement du milieu rural et de l'environnement.

DEA Génie Biologique
Responsable: J F Stoltz (INSERM)
Cohabilitation: UHP; Nancy I; Université de Reims
Objectifs: Formation pluridisciplinaire ayant pour objet la recherche; la conception; la mise au point, l'évaluation et la fabrication de nouvelles technologies d'analyse; de diagnostic et de thérapeutique pour l'instrumentation et les bioréactifs.

DEA Génie des Procédés
Responsable: F Pla
Cohabilitation: –
Objectifs: Formation à et par la Recherche dans les domaines de la chimie physique; des matériaux et du génie des procédés.

DEA Génie des Systèmes Industriels
Responsable: Cl Guidat
Cohabilitation: ENSAI de Strasbourg
Objectifs: Développer l'aptitude à tester et à mettre au point des outils et des méthodes nécessaires à l'analyse et à la compréhension des problèmes de conception; de mise en oeuvre et de gestion des systèmes industriels en évolution technologique.

DEA Géosciences de l'Environnement
Responsable: J Cases
Cohabilitation: Universités d'Aix-Marseille III et Toulouse III
Objectifs: Formation regroupant plusieurs approches cristallographiques; pétrologiques, géochimiques, géophysiques et hydriques dans le domaine de l'environnement continental et marin.

DEA Physique et Chimie de la Terre
Responsable: P Barbey
Cohabilitation: UHP; Nancy I; Université de Strasbourg I
Objectifs: Formation avancée dans les domaines des Sciences de la Terre et des Sciences du Sol orientée soit vers les matières premières; pétrologie et géochimie, soit vers la pédologie; les sols dans les systèmes continentaux.

DEA Informatique
Responsable: M Cl Portmmann
Cohabilitation: UHP; Nancy I; Universités de Nancy II; Metz
Objectifs: Formation avancée en informatique via les concepts de programme d'intelligence artificielle logique; de programmations fonctionnelle et logique et des techniques de communication.

DEA Mécanique et Energétique
Responsable: O Sero Guillaume
Cohabilitation: UHP Nancy I; Université de La Réunion
Objectifs: Assimilation et approfondissement des connaissances théoriques et expérimentales en mécanique et énergétique et ouverture sur des applications de tout niveau dans les secteurs industriels et universitaires.

DEA PAE3S (Protection Aménagement et Exploitation du Sol et du Sous-Sol)
Responsable: F Homand
Cohabilitation: ENGEES
Objectifs: Approndissement des connaissances et développement de recherches dans les différents domaines de l'aménagement et de l'exploitation du sol et du sous-sol; Application aux secteurs spécifiques du génie civil; du génie minier et du génie pétrolier.

DEA Physique et Chimie de la Matière et des Matériaux
Responsable: P Mangin
Cohabilitation: UHP; Nancy I
Objectifs: Conception et élaboration de nouveaux matériaux; étude de leurs propriétés physiques et chimiques en relation avec leur structure et analyse scientifique des phénomènes mis en jeu.

DEA Science et Ingénierie des Matériaux
Responsable: G Lesoult
Cohabilitation: UHP; Nancy I
Objectifs: Génie des procédés d'élaboration, étude de l'influence des méthodes d'élaboration sur les structures obtenues; traitements thermiques et thermomécaniques de matériaux de structure.

DEA Procédés et Traitement de l'Energie Electrique (PROTEE)
Responsable: A Mailfert
Cohabilitation: UHP, Nancy I, Université de Besançon
Objectifs: Formation par la recherche dans le domaine de l'électrotechnique : il s'agit du traitement de l'énergie électrique et ce certains procédés qui l'utilisent. La formation traite donc de machines et d'actionneurs électromagnétiques; d'électronique de puissance; d'électrothermie.

DEA Rayonnements et Imagerie en Médecine
Responsable: R Guiraud
Cohabilitation: Universités de Toulouse III; Paris-Sud; Bordeaux II (Centre Alexis Vautrin)
Objectifs: Développement de l'utilisation médicale des rayonnements.

DEA Sciences de Gestion
Responsable: J L Coujard
Cohabilitation: Université de Nancy II
Objectifs: Formation généraliste; plurithématiques et pluridisciplinaire.

Formation Doctorale en Automatique et Informatique Industrielle de Toulouse

Formation

Le DEA (Diplôme d'Etudes Approfondies) constitue la première année d'étude du cycle de formation à la recherche. Sa préparation se déroule sur une année complète et comprend:

– 160 heures de cours théoriques: Systèmes multidimensionnels; Modélisation et analyse pour la commande des systèmes discrets; Programmation mathématique, graphes et problèmes combinatoires; Planification et ordonnancement; Processus stochastiques; Estimation; Filtrage; Commande optimale; Commande robuste; Systèmes distribués; Evaluation de performances; Robotique etc

– des séminaires de recherche et des séminaires de perfectionnement sur les outils informatiques et les techniques de documentation

– 800 heures (cinq mois temps plein) de stage de recherche en laboratoire universitaire

L'obtention du diplôme est conditionnée par la réussite à un examen sur les cours théoriques et par la validation du stage de recherche qui fait l'objet d'un rapport écrit et d'une soutenance orale devant un jury.

Le diplôme permet de préparer une thèse de doctorat dans un laboratoire de recherche. La durée d'une thèse est généralement de trois ans.

Programmes de Recherche

Le stage de DEA et la préparation de la thèse de doctorat sont effectués dans les six laboratoires d'accueil de la formation doctorale dont certains sont rattachés au Centre National de la Recherche Scientifique (CNRS) et un à l'Office National d'Etudes et de Recherche en Aéronautique (ONERA).

Les domaines de recherche concernent: la Théorie des signaux et des systèmes; l'Automatique des systèmes continus et discrets; les Systèmes de production; le Génie industriel; la Robotique et l'Intelligence artificielle; la Sûreté de fonctionnement; les Réseaux informatiques; l'Algorithmique parallèle; le Génie logiciel; la Commande des machines électriques etc.

Informations Générales

La formation doctorale regroupe quatre établissements toulousains: l'Université Paul Sabatier (UPS); l'Institut National Polytechnique de Toulouse (INPT); l'Institut National des Sciences Appliquées de Toulouse (INSAT) et l'Ecole Nationale Supérieure de l'Aéronautique et de l'Espace (ENSAE).

La formation est ouverte aux titulaires d'une maîtrise ou d'un diplôme d'ingénieur ou d'un diplôme équivalent dans les spécialités de la formation doctorale. L'admission s'effectue sur dossier.

Pour la préparation du doctorat, des allocations de recherche peuvent être attribuées aux ressortissants UEE.

Réussites et Spécificités du Corps Professoral à ce Jour

Toulouse est l'un des plus grands centres français de recherche en Automatique et Informatique. Plus de 150 directeurs de recherche, couvrant un large spectre de spécialités, sont susceptibles d'encadrer des étudiants de la formation doctorale dans les laboratoires d'accueil. Auteurs de nombreuses publications, beaucoup d'entre eux sont membres d'instances nationales et internationales qui définissent les orientations scientifiques et participent à l'organisation de congrès.

Au total 35 enseignants ou chercheurs interviennent dans les cours de la formation doctorale dispensés dans les quatre établissements. Chaque année, sur les 110 doctorants de la formation, une trentaine soutiennent leur thèse.

Ressources et Programmes Distinctifs du Département

Les laboratoires d'accueil sont dotés de moyens informatiques nombreux et performants. Ils participent à plusieurs dizaines de programmes nationaux et internationaux (ESPRIT 4, EUREKA, ERASMUS, TEMPUS etc).

Au travers de conventions d'échanges et de contrats, ils sont en relation avec un grand nombre de partenaires étrangers universitaires et industriels, en particulier Européens.

Environnement Local et Régional

Capitale de la région Midi-Pyrénées, plus grand centre français de constructions aéronautiques, Toulouse est une agglomération urbaine de 600,000 habitants où l'activité culturelle est fortement développée (cinémas, théâtres, concerts), en particulier par la présence de plus de 100,000 étudiants.

Contact

Secrétariat de 3ème cycle EEA
DEA AII
LAAS – CNRS
7 avenue du Colonel Roche
31 077 Toulouse cedex 4
France

Tél (+33) 5 61 33 62 25
Fax (+33) 5 61 55 35 77
Email authie@laas.fr

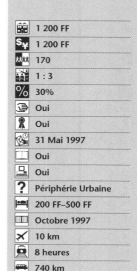

1 200 FF	
1 200 FF	
170	
1 : 3	
30%	
Oui	
Oui	
31 Mai 1997	
Oui	
Oui	
Périphérie Urbaine	
200 FF–500 FF	
Octobre 1997	
10 km	
8 heures	
740 km	

UNIVERSITÉ
TOULON-VAR

Université de Toulon et du Var
Faculté des Sciences et Techniques
DESS Ingénierie Marine, Mention Télécommunications et Robotique

Contact
Professeur Richou
Université de Toulon
UFR Sciences
DESS IM
Faculté des Sciences UTV
BP 132
83 957 La Garde
France

Tél   (+33) 4 94 14 23 18
Fax   (+33) 4 94 14 21 68

Formation

La formation au DESS comporte des enseignements théoriques et pratiques (525 heures) et un stage industriel (cinq mois). Elle débute par des enseignements de mise à niveau dans certaines disciplines (75 heures) pour tenir compte de la diversité des formations préalables des étudiants:

MODULE 1: MÉCANIQUE DES FLUIDES HYDRODYNAMIQUE

MODULE 2: ELECTRONIQUE, AUTOMATIQUE, SIGNAL

Le jury d'admission, au vu des dossiers, répartit les candidats dans ces deux options.

Tous les autres enseignements (430 heures) sont destinés à l'ensemble des étudiants sans choix optionnel:
– Hydrodynamique
– Identification
– Informatique
– Intelligence artificielle
– Instrumentation (robots, systèmes acoustiques, optiques)
– Reconnaissance
– Robotique
– Statistiques
– Traitement du signal
Un certain nombre de conférences (20 heures) complète la formation.

Programmes de Recherche

Les cours ont lieu de Septembre à Février.

Les étudiants effectuent un stage de cinq mois dans un laboratoire de recherche, industriel, ou en entreprise. Il est suivi et évalué par les universitaires et les professionnels. Présentation du mémoire en Juillet.

Informations Générales

Le DESS a pour but de former des cadres du niveau ingénieur.
Dans le secteur de l'Ingénierie Marine: La formation est axée sur les applications des Télécommunications; de la Robotique; de l'Automatique; de l'Intelligence artificielle et des Systèmes experts, en milieu marin.

Cette fonction d'ingénieur pourra être excercée dans les organismes publics et privés dans divers domaines: Etudes; Production; Installation; Exploitation; Maintenance et Gestion de projet.
Dans le secteur de l'Offshore: Il s'agit principalement des domaines de l'Exploration; du Forage; des Supports de travail; de la Production; des Interventions sous-marines et de l'Instrumentation.
Dans le secteur de l'Océanologie: Il s'agit des domaines plus larges tels que l'Electronique; l'Informatique; les Systèmes d'acquisition et de traitement de données.

Conditions d'Inscription

Sont autorisés à s'inscrire les étudiants titulaires d'une Maîtrise de Sciences ou équivalent (diplôme d'ingénieur ou diplôme étranger avec équivalence).

Le jury d'admission prononce en Juillet l'admission définitive, le classement en liste d'attente, ou le refus. En fonction des désistements intervenus parmi les candidats admis, le jury détermine en Septembre ceux des étudiants placés sur liste complémentaire qui peuvent être admis.

🏛	500 FF
💲	500 FF
👥	30 (Dépt)/6 000 (Total)
🏫	–
%	15%
📖	Oui
🏃	Oui
📅	Juillet 1997
📖	Oui
🖥	–
?	Urbain
🛏	600 FF–1 300 FF
🏢	15 Septembre 1997
✈	20 km
🚌	8 heures
🚆	800 km

on the net http://www.editionxii.co.uk

Institut National Polytechnique – ENSEEIHT
Université Paul Sabatier – Toulouse III

INP | **ENSEEIHT**

Formation

DEA DE GÉNIE ELECTRIQUE

Le DEA de Génie Electrique comprend deux spécialités SE et MM (Systèmes Electriques et Matériaux–Matériels).

Le tronc commun comprend 120 heures de cours et chaque spécialité 30 heures de cours.

Une initiation à la recherche et des conférences industrielles viennent compléter les enseignements théoriques.

Une période de mise à niveau de six semaines (du 15 Septembre au 31 Octobre) est proposée avant le début des cours propres du DEA qui se déroulent du 1 Novembre au 31 Mars suivis du stage. La soutenance du stage a lieu en Juin.

Programmes de Recherche

Le DEA de Génie Electrique s'appuie sur plusieurs équipes d'accueil pour les stages (LEEI; LGE; LAAS; CPAT à Toulouse; LEM à Montpellier et ENI de Tarbes ainsi que des Centres de Recherche Industriels).

Tous les thèmes de recherche du Génie Electrique sont concernés:
– Commande des Systèmes Electriques
– Composants
– ELectronique de Puissance
– Machines Electriques
– Matériaux
– Matériels
– Plasmas; etc.

Le stage est commencé à temps partiel entre Novembre et Mars.

Le DEA fait partie d'une formation doctorale qui prend en charge la continuation en thèse de doctorat des étudiants ayant obtenu le DEA avec de bons résultats. Des bourses ministérielles et/ou industrielles peuvent être fournies.

Informations Générales

Pour les élèves ingénieurs, il est possible de suivre le DEA conjointement à la dernière année. De même pour des ingénieurs déjà diplômés le DEA peut être suivi conjointement avec des formations spécialisées (amenant à des diplômes d'ingénieur) au sein de l'INPT. Le stage se prolonge alors jusqu'en Septembre.

Réussites et Spécificités du Corps Professoral à ce Jour

Les enseignants du DEA sont soit des Professeurs ou Maîtres de Conférences, soit des chercheurs du CNRS. Les personnels ont à leur actif de nombreuses publications dans des Congrès internationaux et des revues scientifiques internationales. Des livres et ouvrages (Polycopies) sont disponibles. Au niveau français, un Congrès de Jeunes Chercheurs en Génie Electrique (JCGE) est organisé tous les deux ans.

Ressources et Programmes Distinctifs du Département

Les étudiants du DEA ont accès aux installations d'enseignement et de recherche des établissements. Ils ont accès à des calculateurs. Les laboratoires de recherche accueillent des stagiaires étrangers (communauté et hors communauté) dans le cadre de Programmes Européens ou de collaborations internationales.

Contact
Mme Merlo
INP – ENSEEIHT
Formation Doctorale en Génie Electrique
2 rue Camichel
31 071 Toulouse Cedex
France

Tél   (+33) 5 61 58 83 89
Fax   (+33) 5 61 62 09 76

	1 600 FF
	1 600 FF
	30 environ
	–
%	30 %
	Oui
	Oui
	31 Mai 1997
	–
	–
?	–
	400 FF environ (Ext)
	18 Septembre 1997
	8 km
	–
	730 km

Université Claude Bernard – Lyon I
Ecole Doctorale Matériaux

Contact
Professeur A Hoareau
Ecole Doctorale des Matériaux
Département de Physique
des Matériaux
Université Claude Bernard –
Lyon I
43 boulevard du 11 Novembre
69 622 Villeurbanne cedex
France

Tél   (+33) 4 72 44 85 66
Fax   (+33) 4 72 43 16 08

Formation

DEA Microstructure et Comportement Mécanique et Macroscopique des Matériaux: Génie des Matériaux
Contact:
> Professeur Gérard Guénin
> GEMPPM JNSA
> Bâtiment 502
> 69 622 Villeurbanne cedex
> France
> Tél: (+33) 4 72 43 82 45
> Fax: (+33) 4 72 43 85 28

DEA Matériaux Polymères et Composites
Contact:
> Professeur Henri Sautereau
> LMM INSA
> Bâtiment 403
> 69 621 Villeurbanne cedex
> France
> Tél: (+33) 4 72 43 81 78
> Fax: (+33) 4 72 43 85 27

DEA Sciences des Matériaux et des Surfaces
Contact:
> Professeur Jean-Louis Barrat
> DPM
> Bâtiment 203
> Université Claude Bernard – Lyon I
> 43 boulevard du 11 Novembre
> 69 622 Villeurbanne cedex
> France
> Tél: (+33) 4 72 44 85 65
> Fax: (+33) 4 72 89 74 10

Programmes de Recherche
L'Ecole Doctorale Matériaux de Lyon regroupe trois DEA s'appuyant sur 30 laboratoires et environ 250 chercheurs et enseignants chercheurs. Cette école est actuellement structurée autour de quatre grands thèmes:
– Nanophases; Nanotechnologies
– Matériaux à propriétés spécifiques; Polymères; Alliages métalliques; Céramiques
– Composites et multimatériaux; Matériaux hétérogènes
– Matériaux actifs et capteurs

Informations Générales
– Conditions d'admission dans les DEA: Maîtrise Scientifique; Diplôme d'Ingénieur
– Titre français ou étranger admis en équivalence.
– Sélection des candidats sur dossier
– Pour chaque DEA: Enseignement théorique d'environ 160 heures et stage pratique d'une durée minimale de trois mois dans l'un des laboratoires d'accueil
– Effectif des DEA: 20 à 50 étudiants

Environnement Local et Régional
La ville de Lyon est la deuxième métropole de France et dispose d'un pannel d'activités très varié tant sur le plan culturel qu'artistique.

1 000 FF
1 000 FF
150
–
10%
Oui
Oui
Juin 1997
–
–
–
1 200 FF
Septembre 1997
30 km
2 heures
450 km

Université Pierre et Marie Curie – Paris VI

Ecole Doctorale de Mécanique Fondamentale et Appliquée, d'Energétique et de Robotique

L'Ecole Doctorale comporte quatre formations doctorales et s'adresse aux étudiants relevant de ces formations et inscrits à l'Université Pierre et Marie Curie. Elle rassemble environ 50 laboratoires d'accueil, 180 étudiants de DEA et 200 étudiants en thèse.

Née d'une volonté commune d'harmonisation et de développement concerté, l'Ecole Doctorale a pour objectifs essentiels:

– Offrir aux étudiants le fait qu'elle rassemble une plus grande facilité de dialogue entre les différentes thématiques qu'elle comporte ainsi que des structures centralisées: Association d'élèves et anciens élèves; Fichier d'employeurs; Informations sur les cursus européens et sur la vie scientifique de chacun des DEA

– Mieux faire connaître aux industriels, aux grands organismes de recherche et aux collectivités régionales le potentiel dont elle dispose ainsi que la qualité et les spécifités des quatre formations suivantes qu'elle regroupe:

DEA DE MÉCANIQUE
Responsable: Professeur R Gatignol
Filière 1: Singularités, rupture, comportements non linéaires des solides
Filière 2: Mécanique des fluides: dynamique et systèmes complexes
Filière 3: Systèmes non linéaires, instabilités et turbulence

DEA DES SOLIDES, STRUCTURES ET SYSTÈMES MÉCANIQUES
Responsable: Professeur F Léné
Option 1: Calcul et optimisation des structures mécaniques
Option 2: Mécanique en conception et fabrication
Option 3: Structures, composites et CAO
Option 4: Vibrations des structures
Option 5: Mécanique de la construction
Option 6: Modélisation, calcul et conception
Option 7: Mécanique non linéaire de l'ingénieur

DEA DE CONVERSION DE L'ENERGIE
Responsable: Professeur J Jullien
Filière 1: Ingéniérie des machines à conversion d'énergie
Filière 2: Machines thermiques et combustion
Filière 3: Mécanique des fluides des machines des systèmes propulsifs
Filière 4: Thermique appliquée aux machines

DEA DE ROBOTIQUE
Responsable: Professeur J C Guinot
Option 1: Robotique autonome
Option 2: Ingéniérie des systèmes robotisés
Option 3: Robotique d'intervention et de service
Option 4: Mécatronique des Systèmes Robotisés

Laboratoires d'accueil des doctorants implantés dans les grands organismes de recherche, les écoles d'ingénieurs ou les universités sont les suivants: CEN; ONERA; IFP; CEA; INRIA; EDF; CEMAGREF; Bassin des Carènes; SNECMA; Renault; PSA; SNCF; Ecole Polytechnique; Ecole des Mines de Paris; ENS–Cachan; ENSTA; ENSAM; ENPC; ESCPI; CNAM; ISMCM Saint-Ouen; Universités d'Evry et d'Antilles-Guyane.

Contact
Michèle Larchevêque
Boîte 163
Université Pierre et Marie Curie
4 place Jussieu
75 252 Paris cedex 05
France

Tél   (+33) 1 44 27 35 21
Fax   (+33) 1 44 27 52 59

FRANCE

CENTRALE
P A R I S

Ecole Centrale Paris
Mechanical Engineering
Applied Mechanics

Contact
D Aubry
Ecole Centrale Paris
92 295 Chatenay Malabry
Cedex
France

Tél   (+33) 1 41 13 13 21
Fax   (+33) 1 41 13 14 42
Email aubry@mss.ecp.fr
WWW   ECP.FR

Taught Courses

COMMON CORE
Vibrations of structures; Wave propagation in solids; Random vibrations

NON LINEAR DYNAMICS (MODULE 1)
Computational non-linear dynamics; Dynamic properties of materials; Contact and impact; Multibody dynamics

COUPLING (MODULE 2)
Fluid structure interaction; Aerodynamics; Vibroacoutics; Hydroelasticity; Soil structure interaction

IDENTIFICATION AND CONTROL (MODULE 3)
Structural identification of vibrations, model updating; Inverse problem

Research Programmes
– Structural dynamics including large displacements and deformatures
– Fluid structure interaction
– Contact dynamics
– Adaptive modelling
– Model updating and inverse problems
– Cable mechanics

General Departmental Information
The laboratory of Applied Mechanics at Ecole Centrale hosts about 20 permanent researchers and 40 PhD students. If offers a wide spectra of experimental and computational facilities.

Outstanding Achievements of the Academic Staff to Date
– Special algorithms for algebric differential equation
– Seismic behaviour of dam/valley/reservoir system
– Large displacement and contact of cables
– Wave propagation in unbounded and periodic media
– Metal forming simulation
– Model updating of industrial structures
– Adaptive mesh refinement

Special Departmental Resources and Programmes
Specialisation in the underneath field are proposed by the Laboratory: Structural Mechanics; Mechanics of materials; Soil mechanic.

Outstanding Local Facilities and Features
The city of Paris is very close (20 min by train).

🏛	2 000 FF
💰	2 000 FF
👥	30
🏫	1 : 3
%	5%
💱	No
🏃	–
🕐	July 1997
▭	Yes
💻	Yes
?	University
🛏	–
📅	Sept/October 1997
✈	5 km
🚌	20 mins
🚋	5 km

on the net   http://www.editionxii.co.uk

City University of Hong Kong
Faculty of Science and Technology

香港城市大學
City University
of Hong Kong

Taught Courses
The Faculty has nine Master of Science (MSc) courses which are all designed to provide specialised curricula, encompassing the frontiers of knowledge in selected disciplines and keeping pace with the latest developments in the areas. MSc courses include:
- MSc Applied Mathematics*
- MSc Automation Systems and Management
- MSc Computer Science
- MSc Construction Management
- MSc Electronic Engineering
- MSc Engineering Management
- MSc Environmental Science and Technology
- MSc Materials Technology and Management

*to be offered subject to Senate approval

Research Programmes
The Faculty offers a wide range of supervised research leading to the MPhil and PhD degrees. Research areas include (by departments):

DEPARTMENT OF BIOLOGY AND CHEMISTRY
Environmental Biology and Chemistry; Industrial Biology and Chemistry.

DEPARTMENT OF BUILDING AND CONSTRUCTION
Building Materials and Components; Concretes: Ordinary & Mineral Replacement, Light-weight, Fibre Reinforced, High Performance; Structural Dynamics (of High Rise Buildings); Construction Automation & Robotics; Construction Management: I.T. Applications, Multi-media Teaching Tools, Productivity and Simulation, Quality, Safety; Construction: Economics, Education, Procurement; Fire Dynamics and Safety Modelling; Energy Conservation Management; Intelligent Buildings; Environmental Engineering and Vertical Transport in Buildings.

DEPARTMENT OF COMPUTER SCIENCE
Computer Networking and Distributed Systems; Graphics, Image Analysis and Multi-media and Cognitive Technology.

DEPARTMENT OF ELECTRONIC ENGINEERING
Applied Electromagnetics; Optical Communications and Optoelectronics; Advanced Manufacturing Technology and Semi-conductor Devices; Automation, Industrial Electronics and Systems; Communication Systems; Digital Signal Processing and Data Communications, Teletraffics and Networking.

DEPARTMENT OF MANUFACTURING ENGINEERING AND ENGINEERING MANAGEMENT
Engineering Management; Design and Manufacture and Mechatronics and Automation.

DEPARTMENT OF MATHEMATICS
Asymptotic and Perturbation Methods; Complexity and Computation; Control Theory; Discrete Mathematics; Dynamical Systems; Finite Element Methods; Fluid and Solid Mechanics; Mathematical Modelling; Non-linear Waves; Numerical Analysis; Optimisation; Parallel Processing; Partial Differential Equations; Scientific Computation and Special Functions of Mathematical Physics.

DEPARTMENT OF PHYSICS AND MATERIALS SCIENCE
Atmospheric/Environmental Physics; Laser and Condensed Matter Physics; Materials Science; Radiation Physics; Astrophysics and Instrumentation.

General Departmental Information
The Faculty consists of seven academic Departments and one Faculty Laboratory Centre with more than 200 academic staff. There are 3,700 undergraduates and 550 postgraduate students enrolling in 26 Bachelors and taught Masters degree programmes. In addition, all departments in the Faculty offer MPhil and PhD degree studies. The number of research students currently amounts to about 300.

Outstanding Achievements of the Academic Staff to Date
The Faculty has established a high reputation for the quality of its research work. Academic staff are fully committed to strengthening established research activities in order to develop Centres of Excellence.

Special Departmental Resources and Programmes
The Faculty now has three major research centre funded centrally, viz. Telecommunications Research Centre, Materials Research Centre and the Centre for Environmental Science & Technology. Since early 1996, the Faculty has established Faculty research centres with matching fund from central. These centres include Optoelectronics Research Centre, Centre for Mathematical Science, Centre for Structural Dynamics and Centre for Intelligent Design, Automation & Manufacturing. The strategy of the Faculty is to have each department focus on an area of excellence in order to achieve our mission of becoming a leading university in the Asia-Pacific region in selected areas of science and engineering disciplines.

Contact
Professor Po S Chung
Professor of Electronic Engineering
Dean of Faculty of Science and Technology
City University of Hong Kong
83 Tat Chee Avenue
Kowloon
Hong Kong

Tel   (+852) 2788 7750
Fax   (+852) 2788 7741
Email eeapsc@cityu.edu.hk
WWW
http://www.ee.cityu.edu.hk/fst/pschung.html

Ireland

Cork Regional Technical College
Department of Electrical and Electronics Engineering

Taught Courses

BACHELOR OF ENGINEERING IN ELECTRONIC (FOUR YEARS)
Main subjects:
- Electronic Engineering
- Communication Engineering
- Computer Engineering
- Control Engineering
- Computer Science
- Mathematics

NATIONAL DIPLOMA IN ELECTRONIC ENGINEERING (THREE YEARS)
Main subjects: similar to above taught courses.

NATIONAL CERTIFICATE IN ELECTRONIC ENGINEERING (TWO YEARS)
Main subjects: similar to above taught courses.

Research Programmes

MASTER OF ENGINEERING (TWO YEARS)

DOCTOR OF PHILOSOPHY (THREE YEARS+)
Current research:
- Vision inspection systems
- Adaptive control
- Power control

General Information
The College has 16 departments/schools embracing business; engineering; science; humanities; music and art; and participates in relevant European programmes.

Research takes place in many of the disciplines stated.

Cork, the second city of Ireland, abounds with leisure activities, is near the coast and is the gateway to scenic west Cork.

Application Information
For research programmes: At any time
For undergraduate courses: By 1st February to CAO/CAS, Galway, Ireland

Outstanding Local Facilities and Features
Annual Events:
- International Guinness Jazz Festival
- International Choral and Folk Dance Festival
- Folk Music Festival
- Cork RTC Arts Festival
- International Film Festival

Places:
- Opera House
- City Hall Auditorium
- Cork School of Music Auditorium
- Blarney Castle
- Museums
- Art galleries

Contact
L J M Poland
Department of Electrical and
Electronics Engineering
Cork Regional Technical College
Bishopstown
Cork
Ireland

Tel (+353) 21 326 100
Fax (+353) 21 345 244
Email LUMP@rtc-cork.ie

🚆	IR £900 (year)
$¥	IR £900 (year)
👥	30 (dept)
🏫	1 : 3 (postgraduate)
%	0%
📖	Research Agency Grants
👤	–
🗓	30 June 1997
▢	Extensive
💻	Extensive
?	Multi-site
🛏	£5,100
▥	1 October 1997
✈	6 km
🚌	2 hours 40 mins
🚍	250 km

University College Galway
Department of Engineering Hydrology

Contact
Dr K M O'Conner
Director of Postgraduate Courses
Department of Engineering
Hydrology
University College Galway
Ireland

Tel   (+353) 91 750 341
Fax   (+353) 91 524 913

Taught Courses

I)   HIGHER DIPLOMA IN HYDROLOGY – 10 MONTHS

II)   MSc HYDROLOGY – 15 MONTHS OR MORE

III) MENG SC HYDROLOGY – 18–24 MONTHS

IV) HIGHER DIPLOMA IN ENGINEERING HYDROLOGY – 12 MONTHS

V)   MASTER OF ENGINEERING DESIGN (MED) (HYDROLOGY) – 12 MONTHS

Research Courses

I)   MSc (HYDROLOGY)

II)   MENG SC (HYDROLOGY)

III) PhD (HYDROLOGY)

General Departmental Information
The Department has run International Postgraduate Hydrology courses (MSc and Diploma) since 1979. The majority of the 270 students to date have been from less developed countries. The Department also holds Advanced courses/workshops every second year on River Flow Forecasting.

Outstanding Achievements of the Academic Staff to Date
Academic staff and researchers have published over 20 papers in the *Journal of Hydrology* and other international journals. They have also contributed to the UK Flood Studies Report, a WMO Operational Hydrology Report, and served as Editorial Advisory Board Members and Guest Editor of the *Journal of Hydrology*, and the *Hydrological Sciences Journal*.

Special Departmental Resources and Programmes
Working originally under the leadership of Professor J E Nash, the Department has built a strong reputation in River Flow Forecasting and Modelling and in the area of Stochastic/Statistical Hydrology.

Outstanding Local Facilities and Features
Galway is a small city of about 70,000 inhabitants. It is a regional capital, with good integration between town and University. It is noted for its small theatres, International Arts Festival and friendly atmosphere.

IR £3,500	
IR £8,750	
20 new entrants each year	
–	
80%	
Yes (Selected cases only)	
Yes (Selected cases only)	
1 May 1997	
Yes	
Yes	
Urban	
IR £30–IR £40	
22 September 1997	
10 km	
2.5 hours	
200 km	

Italy

Latvia

Aviation

Riga Aviation University

Riga Aviation University

Taught Courses
- Aircraft and Engine Maintenance
- Computer Aided Design of Electronic Devices
- Computer Complexes and Data Networks
- Computer Transport Information and Management Systems
- Design, Manufacture and Maintenance of Light Aircraft
- Maintenance of Airport Communication Systems
- Maintenance of Aircraft Electric and Navigation Systems
- Maintenance of Aircraft Radio Electronic Equipment
- Metrology, Standardisation and Quality Control
- Motor Car Design, Manufacturing and Testing
- Motor Car Maintenance and Diagnostics
- Transport Engines
- Transport Management
- Transport Repair and Overhaul
- Software for Computer Facilities and Transport Automated Control Systems

Research Programs
Erosion Resistant Coatings; Usage of Renewable Energy Sources; Transport Maintenance Facilities and Technology (Aircraft, ATC equipment, Avionics etc.); Abyssal Subsurface Interface Radar Design; Transport Communication Systems; Navigation Aids Design

General Departmental Information
Two engineering faculties: Faculty of Mechanical Engineering; Radio Electronics and Computer Systems Faculty. These faculties include 13 engineering departments.
The following departments are subordinate to the Faculty of Mechanical:
- Aircraft and Engine Maintenance Department
- Engineering
- Technical Mechanics Department
- Transport Engines Department
- Vehicle Repair and Manufacturing Technology Department
- Vehicle Theory, Calculation and Design Department.
The following departments are subordinate to the Faculty of Radioelectronics and Computer Engineering
- Avionics Maintenance, Metrology and Standardisation Department
- Computer Complexes and Networks Department
- Computer Engineering and Flight Management Systems Department
- Department of Computer Systems for Data Processing and Control
- Department of Radiolocation and Measurement Engineering
- Radio and Electrical Engineering Department
- Software Department
- Telecommunications Department

Outstanding Achievements of Academic Staff to Date
More than 70 percents of the academic staff hold scientific degrees and academic titles, among them 11 are Academician, 179 are Doctors of Science. A number of scientific achievements and dissertation have won Latvia State and international prizes and awards.

Special Departmental Resources and Programs
- Aviation Ground Based Facilities Training Base
- Aircraft Maintenance Training Base equipped with 10 aircraft of various types
- Latvian state Air Traffic Service & RAU Joint Training Canter

Outstanding Local Facilities and Features
University is located almost at the centre of the Riga City. Riga is a capital of Latvia and one of the most attractive places in the Baltics. About 900,000 people inhabit the city. Founded in 1201 Riga is famous for its architecture, historical monuments, theatres and museums. Jurmala, 20 km away from Riga, is popular resort on the Baltic Sea cost. It has been known as a popular spa district since long ago.

Contact
Igor Kabashkin
Riga Aviation University
1 Lomonosova iela Riga
LV-1019
Latvia

Tel   (+371) 710 3215
       (+371) 714 0169
Fax   (+371) 714 0293
Email rau@rau.vernet.lv
WWW
http://www.rau.lv

🖥	$1,200 per year
$¥	$1,700 per year
👥	3,000
👥	1 : 11
%	17%
📖	Yes
👤	–
🕐	1 September 1997
📖	Over 900,000 volumes
💻	Good
?	Self-contained
🛏	$10–$20
🗓	1 September 1997
✈	15 km
🚗	In the Capital
🚌	In the Capital

Lithuania

Vilnius Technical University

Taught Courses
Faculties offer postgraduate courses for the Master of Science degree in: Civil Engineering; Environmental Engineering; Geodesy; Architecture; Automation; Radioelectronics; Information Technology; Technosphere and Ecology; Computer Graphics; Mechanical and Industry Engineering; Building Construction; Transport Vehicle Engineering; Transport Engineering Management; Engineering Management, and Aviation. Doctoral studies are offered in Human Sciences (Architecture); Social Sciences (Economics, Management); Technical Sciences (Agricultural and Environmental Engineering, Electronics and Electrical Engineering, Power Engineering and Heat Engineering, Mechanics Metrology and Measurements, Transport Technology, Building Construction). Foreign students are welcome to take full or part-time studies in separate courses.

Research Programmes
The University offers a wide range of supervised research leading to the PhD degree. Currently active project areas include (by faculties):

Faculty of Architecture – research and design of architectural and urban objects, analysis of failings and renovation of the architectural structure of buildings and their elements, and hall acoustics.

Faculty of Environmental Engineering – harmonisation of urban engineering systems, resources and environment.

Faculty of Civil Engineering – analysis of structures and optimisation theory, development of computational methods, mechanics of reinforced concrete and computer-aided design, reliability of concrete steel, composite, glulam timber structures in design and maintenance stages, automated alternative design of projects and total quality control, increasing quality and durability of building items from local raw materials, soil and foundation interaction, rational types of foundations, noise and vibration, accident prevention and ergonomics.

Faculty of Electronics – Design the optimising of automatic control systems, theory investigation and design of wideband electrodynamics and optoelectronic systems, digital signal processing.

Faculty of Mechanics – design of joint equipment for rational systems, design of new welding power sources and technologies, increase of reliability of welding material and joints, new technologies of metal surface repairing and hardfacing, elastic device, optimum constructions of circular grinding machines, development and updating of mechanical processional systems and their component, tribological process and systems.

Faculty of Transport Engineering – Secure traffic, transport environmental care, transport design with a use of models, reliability of transport machines, logistics in transport, multimodal shipments, international shipments.

Faculty of Business Management – social and economic reforms, culture and practice of management in Lithuania and other states of Eastern Europe, informational management technologies and their development, development of innovation and investment management methods, managerial provision of economical and social reforms, problems of capital market formation, formation of the investment project preparation methodology, formation of economical mechanism for management of ecological environment, research of various ownership types of construction, and production companies activities under market conditions.

Faculty of Fundamental Science – radiation and transport phenomena in materials, environment radioactivity control, gamma-spectrometry application for natural object research, investigation and development of computer geometric and graphic and their employment in design systems, information technology for industrial needs, asymptotic analysis of several statistics and applications of optimisation methods for solving engineering problems, constraining and solving mathematical models using differential equations, analysis of resistance to fracture and optimisation of non-linear structure members, non-traditional methods of computational mechanics for synthesis and control of combined continuum and discrete systems: development and applications, investigation of the resistance to cycle loading or metals and other materials, projection and analysis of optimum asphalt concrete mixture.

General Information
There are eight faculties, a studying institute, two studying centres, consisting of 42 departments, ie. faculties of Architecture, Electronics, Fundamental Sciences, Mechanics, Civil Engineering, Environmental Engineering, Transport Engineering, Business Management, Institute of Aviation, International Studying Centre, Centre for Continuing Education. The University has a library and the publishing house "Technika". The study in the University consists of three stages – BSc, MSc, and doctoral studies – in almost all the fields of engineering and economics, business administration, management, and architecture. There are 4 780 students in BSc degree, 650 in MSc degree , and 94 in doctoral studies. Permanent teaching staff is 659.

Outstanding Achievements of the Academic Staff to Date
There are 383 members of academic staff holding doctorates and habilitated (higher) doctorates.

Outstanding Local Facilities and Features
Vilnius is the capital of the Republic of Lithuania. It is an ancient and very attractive city with many cultural facilities. The university is placed in the wooded suburbs, a few kilometres from the historic city centre.

Contact
Dr Regimentas Ciupaila
Director for International Relations
Vilnius Technical University
11 Sauletekio al.
2054 Vilnius
Lithuania

Tel   (+370) 2 765746
Fax   (+370) 2 700114

US $2,500 (MSc)	
US $3,000 (PhD)	
–	
–	
–	
No	
No	
31 August 1997	
–	
–	
–	
US $100 maximum	
Sept and Oct 1997	
12 km	
–	
–	

LITHUANIA

Kaunas University of Technology

Contact
Assoc. Prof. Petras Barsauskas
Vice-Rector
Kaunas University of
Technology
K. Donelaicio 73
3006 Kaunas
Lithuania

Tel   (+370) 7 22 65 21
Fax   (+370) 7 20 29 12
Email
Petras.Barsauskas@CR.KTU.LT
WWW
http://vm.ktu.lt

🏫	$3,000–$4,000 pa
💲	$3,000–$4,000 pa
👥	–
👨‍🏫	1 : 3.5
%	0.5% : 99.5%
🎓	No
🧍	Requires student exchange
📅	30 June 1997
🛏	Yes
🖥	Yes
?	City centre
🛏	$25–$30 pw
📅	1 September 1997
✈	100 km
�	–
🚌	104 km

Taught Courses
Computer-aided mechanical engineering; mathematical modelling of construction processes; formal specification and analysis for computer network protocols; modelling of computer systems; chemical engineering; food chemistry and technology; public administration, et al, see the internet home page http://vm.ktu.lt

Research Programmes
–   Chemistry: organic chemistry; inorganic chemistry; physical chemistry; polymer chemistry
–   Physics: physical electronics; technical physics
–   Mathematics: applied mathematics; informatics
–   Chemical Engineering: biotechnology; chemical technology; environmental engineering; food chemistry and technology
–   Electronics: radioengineering; electronics engineering technology; telecommunications; ultrasound technology; metrology and measurement
–   Energetics and Electrical Engineering: applied electronics; electrical power engineering; electromechanics; illuminating engineering; process and equipment control
–   Informatics: computer science; software
–   Wood Engineering
–   Textile Engineering
–   Closing and Polymer Products
–   Mechanical Engineering: applied mechanics; machines' production technology; design of machine building equipment; computer-aided design of machines.
–   Transportation Engineering: manufacturing of cars and tractors; design and manufacturing of aircraft
–   Thermal Engineering
–   Materials Technology
–   Civil Engineering: building structure; technology of building materials and structure; technology of building engineering and management; building renovation
–   Management and Business Administration
–   Production Management: management of innovations; business management; management of marketing; commercial management; stock exchanges management; personnel management; international business management; management of banks and financial institutions; finance management
–   Business Adminisration: management; finance; property valuation
–   Public Administration
–   Educology: education management; vocational and business education; organization of higher education; social pedagogy
–   Sociology

General Departmental Information
A person holding a Bachelor's Diploma may be admitted to the Engineering Diploma or Masters programmes. The Engineering Diploma lasts for one year during which an individual collects 40 credits and undertakes a project. Currently every specialist holding the Diploma of Higher Education has the opportunity to apply for the Masters programme. The duration of the Masters degree is two years, during which 80 credits must be acquired and a Master's thesis written.

Any student who completes the Masters degree is eligable to undertake a doctorate. The latter must be completed within five years, two of which must be devoted to studies and research.

In the doctorate programme 40 credits are awarded for studies and 160 credits for research work. The doctoral dissertation is the main qualifying feature of this degree which comprises the synthesis of earlier studies and research in a chosen field. The dissertation should display the doctoral candidate's theoretical and practical ability for independent scientific work. The university awards the diploma is registered by the Lithuanian Council of Science.

Outstanding Achievements of Academic Staff to Date
The university independently publishes monographs, textbooks, and research issues written by its scientific workers and scholars, as well as by scientific staff members of other academic institutions. Eight periodicals have been published in Lithuanian, English, German, French, and Russian, and include summaries in other languages: Chemical Technology; Electronics and Electrical Engineering; Information Technology and Control; Measurements; Mechanics; Materials Technology; Social Science; Environmental Research, Engineering and Management.

The university organises over 30 national and international conferences in the fields of technical, natural and social sciences each year.

Special Departmental Resources and Programmes
The University has several independant research centres, namely: Institute of environmental engineering; Ultrasonic institute; Institute of transport problems; Computer centre; Innovation centre ; Construction reliability centre; Publishing centre; Vibroengineering research centre; Packaging research centre; Centre for Professional development; and KTU-Festo industry automation centre.

Outstanding Local Facilities and Features
Kaunas is the second largest city in Lithuanua, a thriving cultural centre with several theatres, concert halls, a sports hall and museums and art galleries.

The Netherlands

Eindhoven University of Technology
Faculty of Technology Management

tu͡e
Eindhoven
University of Technology

Taught Courses
GRADUATE PROGRAMMES
Two five-year, full-time MSc programmes taught in Dutch:
Industrial Engineering and Management Science
With a main focus on operations management, in particular innovation management, logistics and production control.
Technology and Society
The impact of new technology on society and product design from an ergonomic perspective
  During the last year of their study students have to carry out a major thesis project.
  They can take part in one of the strategic programmes: research projects for a team of 20 undergraduates, who are working together on one specific theme, eg Re-use and Supply Chain Management.

STUDENT EXCHANGE PROGRAMME
The International Student Exchange Programme is a four-month elective course programme in English, from September until December. The Technology Management faculty is involved in several multilateral ERASMUS/SOCRATES networks. Financial grants are available after admission. Maximum 50 students per year.

POSTGRADUATE PROGRAMMES
The postgraduate programme Designers Course in Logistic Control Systems is a two-year, full-time course taught in Dutch and English. Financial grants are available after admission.
  Two four-year full-time PhD programmes taught in English, provided in co-operation with other departments and institutes:
– Graduate School for Business Engineering and Technology Application, focusing on Business Process Design; Operations Management of Business Processes; Information Engineering; and Operations Research and Statistics. Within these fields of expertise the following programmes have been set up: Virtual enterprises; Management of innovation processes and concurrent engineering; and Technology and organisational development
– Graduate School for Perception and Technology, focusing on Sensorial Perception, Language and Speech, Cognition and Communication, and other issues necessary for ergonomic product design

MASTER PROGRAMMES
Several technical MBA programmes provided by the TSM Business School, in which the university takes part, eg the Executive Part-time MBA programme; the Full-time MBA programme; and the Part-time MBA programmes of Project Management, Facility Management, and Business Marketing.
More information:
  TSM Business School
  International Institute for Management of Technology & Business Development
  PO Box 217
  7500 AE Enschede
  The Netherlands
  Tel: (+31) 53 489 80 09; Fax: (+31) 53 489 48 48
  Email: mba@tsm.utwente.nl

Research Programmes
The central theme of the research is management of technology development and technology use. The research, both fundamental and application oriented, focuses on the following fields:
– Innovation Management and Policy
– Operations Management and Logistics
– Perception and Technology

General Departmental Information
The main tasks of the Faculty of Technology Management are education, research and services. The faculty has about 160 academic staff members of which 45 are professors.

Outstanding Local Facilities and Features
Frits Philips Music Centre; Van Abbe Museum for Modern Arts; Yearly Jazz Festival in August.

Contact
Professor dr W van Gelder
Eindhoven University of Technology
Faculty of Technology Management
PO Box 513
5600 MB Eindhoven
The Netherlands

Tel   (+31) 40 247 26 35
Fax   (+31) 40 243 29 83
Email j.f.t.rediker@tm.tue.nl
WWW
http://www.tue.nl/tm

🏫	DFL 2400
💲	DFL 2400
👥	1,900
🏛	1 : 10
%	1 : 30
📖	Exchange Programme
🎓	–
🗓	1 July 1997
🖥	Yes
💻	Yes
?	Urban
🛏	DFL 100
📅	1 September 1997
✈	10 km
🚇	1.5 hrs
🚌	100 km

Poland

Czestochowa Technical University

Taught Courses
The University comprises five faculties with the following course profiles:

FACULTY OF MECHANICAL ENGINEERING
– Mechanics and Machine Building (Masters, BSc and PhD programmes)
– Computer Science (BSc studies)

FACULTY OF METALLURGY AND MATERIALS ENGINEERING
– Metallurgy (Masters, BSc and PhD programmes)
– Materials Engineering (Masters and PhD programmes)

FACULTY OF ELECTRICAL ENGINEERING
– Electrotechnics (masters studies)

FACULTY OF CIVIL AND ENVIRONMENTAL ENGINEERING
– Civil Engineering (Masters studies)
– Environmental Engineering (Masters studies)

INSTITUTE OF MANAGEMENT
– Management and Marketing (Masters studies)
    The language of tuition is Polish, so proficiency in Polish is required. Special Language Training Centres have been established to offer one-year Polish language courses for prospective foreign students.

Research Programmes
Masters graduates are eligible to undertake doctorate studies at our university. The University offers supervised research leading to PhD degree Research topics are negotiated with supervisors on an individual basis. The programme of each student is planned in consultation with his or her main professor and advisory committee. Further details about the research interests of qualified staff members can be obtained from the Deans of Faculties.

General Departmental Information
The University was founded in 1949.
Full-time studies are carried out in two different systems of graduation:
– BSc studies (three and a half years)
– Masters studies (five years)
For the first two to four semesters there is a common core cirriculum for Bachelor and Master degree courses, this followed by separate strands for each degree course.
Students with good results after the first year of their studies may take additional pedagogical training which gives them the right to teach technical subjects in secondary schools.
    Academic year consists of two semesters starting in October and February.

Outstanding Achievements of the Academic Staff to Date
All members of the academic staff are actively involved in research, and numerous books and papers have been published on their respective subjects. Every year they participate in several international conferences.

Special Departmental Resources and Programmes
The University has general research laboratories and specialist research facilities. The main university library has a collection of technical books and scientific periodicals.
    Co-operative agreements and scientific exchange programmes have been established with institutions in many European countries and Japan.
    Research covers many disciplines. Currently active project areas include:
– Applied thermomechanics; Dynamics of mechanical systems; Modern energy generation and energy processing technologies; Applied aerodynamics; Investigation of friction, wear and lubrication of plastic working processes; Non-linear problems in applied mathematics and in mathematical analysis
– Testing, modelling and optimisation of metallurgic processes, metal forming and metallurgic energetics; Metals, alloys, composites, amorphos and magnetic materials testing
– Electrical power; Theoretical and industrial electrotechnics; Scientific description and application of several projects in the field of electrical power, industrial electrotechnics, electronics and electrotechnology
– Clean energy technology; Working out complex industrial sewage-treatment technologies
– Strategic management; Projects of production systems; Logistics management; Modernisation of the theory of accountancy; Projects for the management of regional infrastructure; Development of public sector in market economy; Labour market
    The Student Sport Association (AZS) is one of the most active units at our university. The Physical Training and Sports Centre comprises six different sports sections: Volleyball; Basketball; Football; Martial arts; Tennis and Body building.
Our best sportsmen have successfully participated in many national and international sports events.

Outstanding Local Facilities and Features
Czestochowa is famous for annual pilgrimages to the church and Paulite monastery of Jasna Góra Pilgrims come to the monastery to see the Miraculous Picture of Black Madonna. The spire and the fortifications surrounding the monastery are the symbol of our town. A lot of national and international conferences and seminars are held at Jasna Góra every year. The surroundings of the town are also exceptionally beautiful and are often visited by tourists. There are many hidden caves and grottos as well as old castles and castle ruins sited on craggy hills.

Contact
Olga Stawska
Foreign Relations Office
Czestochowa Technical University
Politechnika Czestochowska
UL. Dabrowskiego 69
42-201 Czestochowa
Poland

Tel (+48) 34 61 28 55
(+48) 34 25 02 52
(+48) 34 6125 80
Fax (+48) 34 612385
Email dn@matinf.pcz.czest.pl
WWW
http://www.matinf.pcz.czest.pl
http://www.k2.pcz.czest.pl

No fees	
US $2,000–US $4,000 pa	
9,380	
1 : 16	
–	
Limited	
Yes	
30 June 1997	
Yes	
Yes	
Multi-site	
US $10–US $20 pw	
1 October 1997	
40 miles	
3 hours	
180 miles	

Romania

General Engineering

Romania

University "Petrol-Gaze" of Ploiesti
Petroleum & Gas Institute

Taught Courses (specialisations)

STUDIES FOR THE MASTER DEGREE
- 1. Deviated Well Drilling (one year, MEng)
- 2. Petroleum Reservoir Engineering (one year, MEng)
- 3. Crude Oil and Gas Production (one year, MEng)
- 4. Off-shore Oil Field Equipment (one year, MEng)
- 5. Reliability of Petroleum Equipment (one year, MEng)
- 6. Refinery and Petrochemical Equipment
- 7. Modern Petroleum Technology and Petrochemisty (one year, MEng)

THE SCHOOL OF POSTGRADUATE ACADEMIC STUDIES
- 1. Petroleum Reservoir Engineering - Waterflooding (two years, MSc, in English)
- 2. The Economic Development of Enterprises (2.5 years)

Research Programmes
The University "Petrol-Gaze" Ploiesti offers facilities for research in: Petroleum Engineering, Mechanical and Electrical Engineering; Petroleum Technology and Petrochemistry.
   Postgraduates are accepted to study for PhD degrees by research.
Examples of research programs: Petroleum and Gas Drilling and Production; Hydraulics and Fluid Mechanics; Petroleum Equipment; Tribology; Control Systems; Strength of Materials; Organic Chemistry; Engineering of Chemical Processes; Petroleum Technology and Petrochemistry.

General Departmental Information
Petroleum and Gas Institute has three faculties: Well Drilling and Petroleum Reservoir Engineering; Mechanical and Electrical Engineering; Petroleum Technology and Petrochemistry.
   The main departments of this faculties are: Drilling – Production; Hydraulic and Reservoir Engineering; Geophysics – Geology; Physics; Petroleum and Petrochemical Equipment; General Mechanical Disciplines; Electrotechnics, Electronics and Automatics; Automatics and Computers; Chemistry; Chemical and Petrochemical Engineering, Petroleum Processing Engineering.

Outstanding Achievements of the Academic Staff to Date
Specific papers, books, courses published for engineering and students.
Special research activities, e.g: well deletion and directional drilling; Increase of recovery final factor; Hydraulics of petroleum and gas recovery; Construction, manufacturing and quality insurance of drilling – production petroleum equipment; Tribology of petroleum equipment; Advanced process control of chemical and petrochemical plants; Engineering of chemical reactions; Phase equilibrium; catalysis and catalysts; Environmental protection in hydrocarbon processing industry.

Special Departmental Resources and Programmes
- Research Activity in co-operation with industry
- Funds from Ministry of Education (budget)
- Sponsoring firms provide - financial support for their students
- EU financial support (e.g. TEMPUS/Phare programme)
- Microproduction and services in the University workshops

Outstanding Local Facilities and Features
- "Toma Caragiu" State Theatre
- Philarmonic orchestra of Ploiesti
- Petroleum Museum; Clock's Museum; Arts Museum
- Touristic places: Prahova Valey; Teleajen Valey, Bucegi Mountains; Ciucas Mountains

Contact
Prof. dr. Niculae N Antonescu
University "Petrol-Gaze" Ploiesti
39 Bucharest Avenue
2000 Ploiesti
Romania

Tel   (+40) 44 17 31 71
Fax   (+40) 44 11 98 47
Email nanto@csd.univ.ploiesti.ro

🏫	–
💲	$340 / month
👥	310
🏢	100/310
%	2%
🎓	70 (for MEng students)
🏃	–
📅	September 1997
☐	Yes
💻	Yes
?	Urbain
🛏	–
📅	1 October 1997
✈	40 km
🚂	1 hour
🚌	60 km

Russia

Novosibirsk State Academy of Civil Engineering
Department of Civil Engineering

Taught Courses
MSc – TWO SEMESTERS (16–18 WEEKS IN SEMESTER), BASED ON EIGHT SEMESTERS BSc COURSES
The two main directions of programmes leading to Msc degree are:
1. Civil Engineering, including the following special programmes:
– Environmental Pollution
– Heating, Ventilation and Air-Conditioning
– Hydraulic-Engineering Structures
– Soils and Foundations
– Structural Materials
– Structures (namely: Metal Structures, Concrete Structures, Structures of Wood and Plastics)
– Water Supply and Water Quality Control
2. Civil Engineering Management and Civil Engineering Economics and Planning. Each programme includes 10 – 12 courses on average, both basic and special.

Research Programmes
DPH AND DSC SIX SEMESTERS (THREE ACADEMIC YEARS)
The main types of programmes leading to DPh and DSc degrees (in accordance with scientific specialities) are:
– Architecture (Theory and History, City Planning, Landscape Architecture)
– Civil Engineering Economics and Planning
– Heating, Ventilation, Air-Conditioning, Gas Supply
– Humanitarian and Social Sciences (Economics, Sociology, Philosophy, Russian History).
– Hydraulic-Engineering Structures
– Mechanics of Fluids, Gas and Plasma
– Soils and Foundations
– Structural Materials and Products
– Structural Mechanics and Strength of Materials
– Structures (Metal, Concrete, Wooden and other types of Structures)
– Surveying and Mapping
– Technology and Managment in Civil Engineering
– Water Supply and Water Resourses Protection

General Information
The language of instruction is Russian. The academic year consists of two semesters starting in September and February. Novosibirsk Academy of Civil Engineering is the oldest and largest higher educational establishment specialised in its sphere in the Asian part of Russia. It is also one of the leaders among similar institutions.

Outstanding Achievements of the Academic Staff to Date
53% of academics have the DPh or DSc degrees or are professors. 13 of them are members of Russian Scientific Academies.
    The State Academy is the publisher of the Russian scientific and technical journal: *Civil Engineering*.

Outstanding Local Facilities and Features
Novosibirsk is the largest industrial, academic, and cultural centre in the Asian part of Russia. The Regional Departments comprise four all-Russian Academies of Sciences. The State Public Scientific and Technical Library has nearly 10 million volumes. Over 70 000 students are in higher educational institutions. Many international conferences in different disciplines are held permanently. The cultural requirements can be satisfied by the opera and ballet theatres (named "Bolshoy Theatre of Siberia"), three professional drama theatres, an excellent Philharmonic Symphony Orchestra, numerous pop and rock music festivals, etc. The sport facilites are both on the Civil Engineering Academy of inter-institutions and an all city level provides wide opportunities in sporting activity.

Contact
Arkadiy P Yanenko
Vladimir G Sebeshev
Vice Rector for Scientific Work
Professor
Novosibirsk State Academy
of Civil Engineering
Department Civil Engineering
113 Leningdradsky Street
Novosibirsk
RF 630008
Russia

Tel   (+7) 3832 660739
Fax   (+7) 3832 660939
Email uungas@ngas.nsk.su

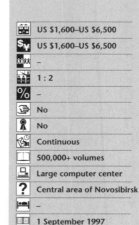

🏛	US $1,600–US $6,500
💲	US $1,600–US $6,500
👥	–
📊	1 : 2
%	–
🔖	No
🎓	No
🕐	Continuous
📚	500,000+ volumes
🖥	Large computer center
?	Central area of Novosibirsk
🛏	–
📅	1 September 1997
✈	30 km
🚉	48 hours
🚌	3,300 km

Rostov-on-Don State Building Academy
Construction

Contact
Shumeiyko Victor-Rector
Rostov-on-Don State Building
Academy ul. Socialisticheskaja
162
344022 Rostov-on-Don
Russian Federation

Tel (+7) 8632 64 45 77
Fax (+7) 8632 65 57 31

Taught Courses
MSc COURSES – TWO YEARS (AFTER BSc)
D ENGINEER – FIVE YEARS
– Industrial and Housing Construction
– Highway and Air Field
– Economics and Management at Enterprises/Building
– Transportation, Organisation and Transport Management
– Bookkeeping and Auditing
– Design and Calculating of Building Structures
– Mechanics of Solids and Theory of Engineering Structures
– Organisation and Building Management
– Mechanisation and Automatisation of Building
– Technology of Building Materials and Structures
– Concrete and Reinforced Concrete Technology
– Water Supply and Water Branching
– Heat and Gas Supply and Ventilation
– Applied Geology

Research Programmes
Postgraduate and Doctorate Courses, Scientist and Pedagogical Study.
PhD – THREE YEARS (AFTER MSc)
DSc – THREE YEARS (AFTER PhD)
– Structures and Constructions
– Building Materials and Units
– Building Mechanics
– Labour and Fire Safety
– Footings and Foundations
– Heat Supply, Ventilation, Air Conditioning, Gas and Light Supply
– Water Supply, Sewage, Building Systems of Water Reservoir Safety
– Technology and Organisation of Industrial and Housing Construction
– Geodesy
– Economics Planning Organisation, Economy Management and its Branches
– Road and Building Machinery
– Industrial Transport
– Highway and Air Field Construction
– Theoretical Fundamentals of Thermotechnics
– Thermal and Molecular Physics
– Mechanics of the deformed Solid Body
– Mathematical Analysis
– Physics of Solids
– Nonorganic Chemistry
– Physical Chemistry
– Russian Language
– Theory of Probability and Mathematical Statistics
– Mathematical Cybernetics
– Geophysics
– Engineering Geology, Froze Study and Soil Study
– Mechanical Systems Management
– Space surveying, Photogrammetry and Photopography

General Departmental Information
The language of tuition is Russian. The academic year consists of two semesters and runs from September to July. The Academy offers a foundation course in Russian Language.
The Academy is the leading Higher Educational Institution in the field of Civil Engineering. It is the centre of science and education in the southern region of Russia.
PhD and DSc degrees are offered. The Academy trains foreign students for 40 years. About 250 foreign students study at the Academy today.

Outstanding Achievements of the Academic Staff to Date
There are 35 chairs at the Academy. More than 75% of the teaching staff with PhD and DSc degrees are working at the Academy now. Among them are 60 DSc; professors, seven Honoured Scientists; four Members; five corresponding members and two councillors of the Academy of Science of Russian Federate Republic.
For the last five years 21 monographics; nine text-books; 103 teaching aids; 40 collections of scientific works and more than 1,250 scientific articles were published. The Academy scientists carry out 16 themes in the programme of State Committee of Higher Education "Architecture and Designing", 13 themes on competition on grants.

Outstanding Local Facilities and Features
Rostov-on-Don, the centre of the southern region of Russia. The population is over 11,250,000 people. Rostov is a big scientific, cultural and industrial centre. There are over 50,000 students in Rostov. 15 Institutes and University are situated in Rostov.

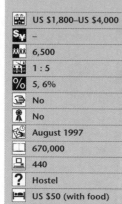

🏫	US $1,800–US $4,000
💲¥	–
👥	6,500
⚥	1 : 5
%	5, 6%
📖	No
🧑	No
📅	August 1997
📖	670,000
💻	440
?	Hostel
🍽	US $50 (with food)
📅	1 September 1997
✈	8 km
🕐	22 hours
🚌	1,100 km

Astrakhan State Technical University

Contact
Professor Yuri N Kagakov
Astrakhan State Technical
University
16 Tatischeva Street
Astrakhan
414025
Russia

Tel   (+7) 851 2 25 09 23
Fax   (+7) 851 2 25 64 27
Email
postmaster@atif.astrakhan.su

Taught Courses

MSc/M ENG (TWO YEARS AFTER BSC)

PG DIPLOMA OF ENGINEERING (FIVE YEARS AFTER SECONDARY SCHOOL)
The University's five Faculty Departments offer the following specialisation:
- *Faculty of Fisheries*: Commercial Fishing; Water Bioresources and Aquaculture; Fish and Fish Products Processing; Chemical Processing of Natural Energy Carriers and Carbon Substances; Complex Utilisation and Protection of Water Resources; Fishery; Nature Development; Chemical Engineering and Biotechnology; Food Processing
- *Faculty of Automatics and Computer Technology*: Automation of Engineering Processes and Manufacture; Automated Systems of Information Processing and Management; Automation and Management; Computer Science and Technology
- *Mechanical Faculty*: Machinery and Apparatus of Food Manufacture; Mechanisation of Cargo Handling; Low Temperature Physics and Engineering; Production Machinery and Equipment; Land Transport Systems; Power-Plant Engineering
- *Ship Power Engineering Faculty*: Ship Power Plants; Ship Power Plant Operation; Ship Electrical and Automatic Equipment; Shipbuilding; Electrical Engineering, Electromechanics and Electrotechnics; Operation of Transport Facilities; Ship Building and Ocean Engineering
- *Faculty of Economics and Law*: Finance and Credit; Accounting and Audit; Economics and Business Management (in different branches of industry); Jurisprudence; Management; Economics

Research Programmes

ASTU offers research programmes leading to PhD by dissertation in the following fields: Dynamics and Strength Ability of Machinery; Apparatus and Instruments (Mechanical Engineering; Ship Power Plants: Main and Secondary Elements; Machinery and Apparatus in Refrigerating and Cryogenic Engineering and Air-Conditioning); Lifting- and Conveying Machines; Components and Devices of Computing Facilities and Control Systems (Electrical Complexes and Systems, Operation and Control, Computer-Aided Design Systems); Environmental Protection and Natural Resources Rational Utilisation (Ichthyology, Biotechnology); Economics and Administration of National Economy.

All research programmes are available on three year full-time or four year part-time options.

PhD entry requirements are an MSc, M Eng or its equivalent in engineering or science (e.g. PG Diploma of Engineer).

General Departmental Information

Astrakhan State Technical University was founded in 1930 as Astrakhan Technical Institute of Fisheries and grew into one of the largest higher education institutions in South Russia. The University is situated in a beautiful green area next to the centre of the town, and its modern campus provides a very agreeable working and recreational environments. The University possesses five teaching buildings, a computing centre, three libraries, nine halls for residence, a roofed swimming pool, a stadium, cafeterias, a medical-aid centre and a club-house.

Over 3,000 graduate and post-graduate students from Russia and abroad are trained here. The staff of 35 academic departments comprises 22 academicians, 36 full professors, 210 associate professors as well as lecturers, instructors and support personnel.

ASTU is involved in the Federal Research Programme "Ecological Safety of Russia" and the international research programme under INTAS "Climatic Precipitation and Lake Bottom Sediments in south-east Russia and Kazakhstan".

Outstanding Achievements of the Academic Staff to Date

The majority of staff are engaged in research activities with immense output of papers, dissertations and books. The University interacts with regional branches of industry and contributes a lot to the development of local business community. The University also plays host to various scientific seminars, conferences and congresses of national and international scope.

Outstanding Local Facilities and Features

Regional Museum; Art Gallery after Kustodiev; four local theatres; musical festivals; sport contests; boat-trips to the State Natural Reserve in the Volga-river delta.

🏫	–
💲	US $6,500–US $7,500
👥	110
🏫	1 : 5
%	1 : 22
✈	–
🚶	–
	30 September 1997
📖	457,550 volumes
🖥	Yes
?	Modern
🍽	US $3–US $25
📅	1 November 1997
✈	15 km
🚆	29 hours
🚗	1,500 km

Moscow State Aviation Institute (MAI)
Technical University

Contact
Mr A Kalliopin
Vice President for
International Affairs
Volokolamskoe shosse 4
125871 Moscow
Russia

Tel   (+7) 095 158 0465
Fax   (+7) 095 158 2977
Email
inter@intdep.mainet.msk.su

Taught Courses
MSc – FIVE AND A HALF YEARS

AIRPLANE AND HELICOPTER DESIGN
Provides experience in design, including international projects.

AERODYNAMICS
Provides specialised training in Aerodynamic analysis; Design; Experimental and research techniques for all aircraft.

AIRCRAFT DYNAMICS AND FLIGHT CONTROL
Introduces a study of the behaviour of all types of flying vehicles, including Airplanes; Helicopters; Guided missiles.

AIRCRAFT AND SPACECRAFT ENGINES
Solid fuel; Liquid fuel; Plasma and other engines, at the student's choice.

SPACECRAFT POWER PLANTS, ENERGY AND ENERGOPHYSICAL INSTALLATIONS
Provides theoretical and practical experience in the elaboration of installations and power plants applied in aerospace technology.

AIRCRAFT AUTOMATICS CONTROL SYSTEMS
Advanced training in the design of devices and complexes for aircraft control.

AIRCRAFT INSTRUMENTS AND INFORMATION
Prepares specialists in Measuring instrument design; Communication means; etc.

AIRCRAFT ORIENTATION AND NAVIGATION
Specialises in the field of flight simulation and aircraft guidance.

AVIONICS
Embraces the fields of Radiolocation; Data transmission; Radioelectronic equipment design, antenna-feeder devices construction.

AIRCRAFT INDUSTRY ECONOMICS
Prepares graduates for careers at senior level in aircraft production management, including marketing and public relations.

MANAGEMENT INFORMATION SYSTEMS
Computer application systems and networks to industrial and trade management.

SPACECRAFT DESIGN AND CONSTRUCTION
Specialises in the design of all types of satellites, launchers, and other spacecraft.

LIFE AND FLIGHT SAFETY SUPPORT SYSTEMS
Trains specialists in the safety of orbital complexes.

APPLIED MATHEMATICS AND PHYSICS
Provides training and research programmes in Quantum mechanics; Software elaboration; and other fields of exact sciences applied in aerospace technology.

Research Programme
Well qualified graduates taking MSc degree may register for three year PhD programmes in any of the above fields.

General Information
Moscow State Aviation Institute is one of the largest educational and research facilities in Russia. It was founded in 1930 with three colleges with 300 students. It has now nine basic colleges, six branches and about 25,000 students who can choose among 30 basic aerospace specialities, plus non-aerospace related ones. The institute offers great opportunities for research activities using 46 main laboratories and 80 research departments. The network of Students' Design Bureau allow the students to develop their scientific potential.

Outstanding Local Facilities and Features
Moscow has many theatres, museums, etc.

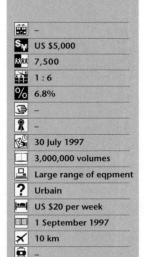

🏫	–
💲	US $5,000
👥	7,500
👥	1 : 6
%	6.8%
📖	–
🚶	–
📅	30 July 1997
📚	3,000,000 volumes
💻	Large range of eqpment
?	Urbain
🛏	US $20 per week
📆	1 September 1997
✈	10 km
🚇	–
🚌	–

RUSSIA

Moscow State Academy of Food Production

Taught Courses

MAIN TYPE OF PROGRAMMES OFFERED: BSc, MSc

Food Technology; Chemistry and Biotechnology; Thermoengineering; Automation and Control; Economics; Commerce; Management; Environmental Protection; Information Systems in Economics; Technological Machines and Equipment; Information Technology and Computer Engineering; Metrology, Standardization and Certification.

DIPLOMA ENGINEERING

TECHNOLOGY

Grain storage and processing; Mixed fodder; Breadmaking, confectionery and pasta goods; Technology of canning and food concentrates; Fruit and vegetables processing; Fermentation and wine making; Sugar substances; Fat processing; Ester oil, synthetic fragrances; Perfume cosmetic products; Standardisation and certification of foods.

BIOTECHNOLOGY

Proteins and biologically active substances; Bioconversion of vegetable raw materials; Designing of biotechnological production enterprises; Chemical technology of synthetic biologically active substances.

INSTRUMENT ENGINEERING

Metrology and metrological support.

ECONOMICS

Commerce; Economics and control in the branches of the agricultural-industrial complex; Organisation and planning of manufacturing and labour resources; Business management and commodity and stock exchange activity; Organisation and legal support of the commercial activity of enterprises; Accounting; Analysis and control of economic activity; Management and marketing; Bookkeeping and auditing, information systems in economics.

HEALTH AND SAFETY

Health and safety; The engineering of environmental protection in the food industry; Safety of technological processes and production safety.

AUTOMATION AND CONTROL

Automation of food technology processes and enterprises; Automation of analytical control of food technology processes.

MECHANICAL ENGINEERING IN THE FOOD INDUSTRY

Machinery for the food industry; Design and construction of equipment for the food and grain processing industries; Thermoengineering technologies; Service.

INFORMATION TECHNOLOGY AND COMPUTER ENGINEERING

Automated systems of information processing and control.

Research Programmes

The Academy is eager to inform all the interested establishments that it is ready to work in collaboration with them according to the following directions of its activies:
– Improvement of biotechnological processes and creation of new kinds of equipment for the production of enzymes, protein substances, vitamins and different improvers of food quality
– Improvement of technology and equipment for sugar, fats and oils, bread making, confectionary, pasta and wine making industries
– Improvement of technology and equipment for grain milling, elevators and feeds industries
– Automation of technological processes, application of robototechnical systems and creation of flexible continuous lines
– Improvement of technological processes control methods and foods and raw materials qualities
– Problems of environment protection and human ecology
– Problems of heat technology energy in food industries
– Improvement of technology, equipment for drying and storage of plant raw materials
  The forms of proposed cooperation are the following joint research: Creation and mastering of new technologies; Equipment and instruments, Colloboration in the field of personal training.

General Departmental Information

The Academy is a prominent research, educational and manufacturing centre of Russian Federation. The Academy initiates and implements large-scale scientific, technological and social programmes as well as projects concerning the development of the food processing industries. It also coordinates work with branch and academic organizations. Programmes leading to Science Degree (PhD; DSc) are available in five research fields at the Depatment of Postgraduate Studies. Specialized Academic Boards award Doctor and Candidate of Science the Degrees in 14 specialities. The graduates who would like to obtain a PhD (in Technology, Biology, Chemistry Mechanical Engineering, and economics) are offered postgraduate courses following individual programmes.

Outstanding Achievements of the Academy Staff to Date

MAIN FACULTIES

The Faculty of Technology and Production Management; The Faculty of Production Equipment and Thermoengineering; The Faculty of Cybernetics and Automation of Food Production; The Faculty of Economics and Bussines Management; The Part-time Education Centre; The Faculty of Humanities and Social Sciences; The Admissions Tutorial Centre.

Contact
Foreign Relations Department
Professor Evgeny A Prokofyev
Moscow State Academy
of Food Production
11 Volokolamskoe shosse
Moscow 125080
Russia

Tel   (+7) 95 158 7159
Fax   (+7) 95 158 0371
Email VIT@mgapp.msk.su

US $2,500 per year
US $1,500–US $3,000
5,000
1 : 8
5
–
–
25 August 1997
930,000
120
Dormitory
US $25–US $30 pw
1 September 1997
40 km
–
–

RUSSIA

Chelyabinsk State Technical University
Engineering

Contact
Dmitry Sherbakov, PhD
Head of International
Affairs Department
Chelyabinsk State Technical
University
76 Lenin Prospect
Chelyabinsk 454080
Russian Federation

Tel   (+7) 3512 656 504
Fax   (+7) 3512 347 408
Email dgsh@inter.tu-chel.ac.ru
WWW
http://www.tu-chel.ac.ru/

Taught Courses
Undergraduate engineering departments at CSTU: Automatic-Mechanical Engineering; Automobile and Tractor Engineering; Architecture and Engineering; Aerospace Engineering; Metallurgical, Mechanical Technological Engineering; Apparatus-Building; Applied Mathematics and Physics; Light Industry and Services; Power Engineering.

Length of study is five–five and a half years, resulting in the professional degree of engineer.

Research Programmes
PhD – three years, DSc – five years: Mathematical Analysis; Mathematical Logic/Algebra/Integer Theory; Machine Dynamics and Durability-Devices and Apparatus; Radio Physics; Laser Physics; Physical Chemistry; Automobiles and Automobile Detailing; Machine-Building Technology; Methods of Control and Diagnostics in Machine-Building; Metal Pressure Processing and Finishing; Welding Machines and Technology; Thermal Engines; Hydraulic Machines; Ground Complexes; Electrical Mechanical Engineering; Electrical Engineering Complexes; Semiconducting Transformers of Electrical Energy; Control in Technical Systems; Automation of Technological Processes and Production; Systems of Processing Information; Electrical Power Stations and Networks; Industrial Thermal Energy; Physics of Metals and the Heat Treatment of Metals; Metallurgy of Ferrous/Non-Ferrous/Rare Metals; Foundry Production; Metal Rolling; Exploitation of Automobile Transport; Building Design; Building Materials and Products; Technology and Organisation of Industrial and Civil Engineering; Building Mechanics; Safety in the Workplace.

General Institution Information
Chelyabinsk State Technical University was founded in 1943. 117 DSc's teach full-time. All classes are taught in Russian. Foreign students can take preparatory courses in Russian language and literature. Over the past three years, 86 students from the United States, Canada, and Ireland have enrolled in CSTU's prep courses.

The academic year, which consists of two semesters, begins in September and ends in June.

Outstanding Achievements of the Academic Staff to Date
Over the past five years, professors at the university have published 23 books (in construction engineering, ferrous metallurgy, physical metallurgy, metal pressure processing technology, etc.), more than 900 textbooks have been published, and 200 dissertations have been defended. The academic staff publish in European, Russian and American scientific journals as well as attend world conferences in their respective fields.

Special Departmental Resources and Programmes
The university library is able to access all libraries via Internet. Chelyabinsk State Technical University has 19 computer laboratories which contain over 1,000 computers. CSTU staff and students have access to and regularly use Internet. The Engineering Department publishes two electronic mail international scientific journals. There is a laboratory for researching metal pressure processing technology, for manufacturing and studying metal glasses, and for testing and assembling computers. Due to the fact that Chelyabinsk is an industrial city, scientists work closely with factories, and do a great deal of research there. Annual international conferences are held at CSTU in the fields of mathematics, physics, and economics.

Outstanding Local Features and Facilities
Population 1.1 million. There are many theatres and art galleries as well as a professional hockey team in Chelyabinsk. Chelyabinsk is located in the southern Urals and there are numerous vacation spots in the Urals and surrounding lakes with cross-country and downhill skiing, swimming and boating facilities.

US $900 per year	
US $1,300–US $2,000 py	
12,500	
1 : 9	
–	
Yes	
–	
20 March 1997	
Yes (2 million volumes)	
Yes	
Urban	
US $20–US $50	
1 September 1997	
20 km	
36 hours	
2 000 km	

Far-Eastern State Technical University

Department of Science (Postgraduate Education)

Research Programmes

Mathematics; Mechanics; Physics; Geology and Mineral Science; Mechanical Engineering and Machine-Building; Construction Materials Treatment; Naval Architecture and Shipbuilding; Electric Engineering; Information Science; Computers, Control and Automation; Electric Power Engineering; Minerals Exploration; Powder Metallurgy; Civil Engineering; Safety of Life Activities; History; Economics; Philosophy; Pedagogic; Architecture; Sociology; Political Science.

General Information

Vladivostok is situated at the southern part of the Russian Far East on the coast of the Sea of Japan. The city is the capital of Primorskiy krai. The population is about 800,000. There are 11 higher educational institutions in the city. The Far-Eastern State Technical University is one of the biggest universities in Siberia and the Russian Far East. The University facilities (library, laboratories, etc.) provide good research opportunities.

Outstanding Achievements of the Academic Staff to Date

More than 60 professors are members of the Russian Academy of Sciences and different branch academies.
–   Professor Yuri N Kulchin is a member of the editorial boards of a number of scientific magazines: Laser Biology (China); Technical Ecology (USA). He was also a member of the organising committee of the International Symposium on Laser Biology '96
–   Professor Alexander T Bekker is a member of the ISOPE Conference Programme Committee (USA)

Special Departmental Resources and Programmes

Doctor of Science Degree Courses (Three Years After PhD or Candidate of Science)

Naval Architecture and Shipbuilding; History of Science and Engineering; Safety of Life Activities.

Second Higher Education (Two Years)

Economics; Linguistics; Environmental Protection; Social Work; Public Relations; Computer Science.

Re-Training Programmes

Office Management and Management for Unemployed; Re-Training of retired military officers.

External Programme

University of London Diploma in Computing and Information Systems.

Internationa Congress

Every odd year, in April, students and post-graduate students may take part in the International Students' Congress of the Asian-Pacific Region Countries.

Contact
Andrei M Uroda
Foreign Affairs Office
Far-Eastern State Technical University
10 Pushkinskaya St
Vladivostok
690600
Russia

Tel   (+7) 4232 268 769
Fax   (+7) 4232 266 988
Email root@dpicnit.marine.su

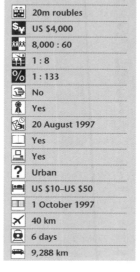

20m roubles
US $4,000
8,000 : 60
1 : 8
1 : 133
No
Yes
20 August 1997
Yes
Yes
Urban
US $10–US $50
1 October 1997
40 km
6 days
9,288 km

Irkutsk State Technical University

Contact
V S B Leonov
Irkutsk State Technical
University
Irkutsk Lermontov Str. 85
Russia 664074

Tel   (+7) 3952 43 16 12
      (+7) 3952 43 16 42
Fax   (+7) 3952 43 05 83
Email oms.ipi@irkutsk.su

Taught Courses
MSc – two semesters (16 – 18 weeks in semester), based on eight semesters BSc courses. The technical exploitation of flying apparatus and engines; lifting transport and road building machines and equipment; motor cars and automotive economics; civil engineering; road and airport construction; architecture; geological survey, search and prospecting; hydrogeology and engineering geology; technology and technical means for prospecting useful mineral deposits; mine surveying; underground mining of minerals; opencast mining; mining machine equipment; engineering geodesy; automatisation and electrification of mining works processes; automatic control system; computers and computing systems; economic information and ASU; radiotechnology; technology of machine-building; metalcutting machine tools; ore dressing; physico-chemical researches of metallurgical processes; chemical technology of organic materials; the chemical technology of non-metal and cilicate materials; communal building; the production of building materials and construction; heat and gas supply; ventilation air base in protection; water supply, sewerage systems, rational use of water resources; electrical station; electric power supply of industrial enterprises, town and agriculture; thermal power stations; electric transportation; power engineering; electrical drive and automatisation of industrial installation; law; social pedagogic; sociology; world economics; design.

Research Programmes
Candidate of science – three years full-time; four years part-time; DSc – three years.
Mining machines; underground mining; physical chemistry; exploitation of automobile transport; organic chemistry; geophysical methods of prospecting; ore dressing; hydrogeology.

General Departmental Information
The Irkutsk State Technical University was founded in 1930, in 1960 it was turned into Polytechnical and in 1993 it was reorganised into State Technical University. The University has trained foreign students since 1961. It is interested in attracting foreign students for studies and inviting foreign teachers to deliver lectures as well as in establishing links with foreign higher educational institutions.

There is an International Department at the University. The International Department has modern equipment, lingual and computer classes. The Russian and other courses are taught there. The successful finishing the international department gives the right to study at institutes of Russia and to receive the specialities of engineering, economic – engineering, medical and philological profiles. There are also summer intensive courses of Russian language at the international department. The courses are conducted for students of any nationality. Their major goal is to develop intensely the skills to converse. Seminars devoted to Russian grammar, economy and literature are offered for your choice. For specialists intending to work in Russia the courses contemplate mastering the language with regard to the business sphere, i.e. ability to be involved and speak fluently in Russian, to read, to interpret, to listen and understand, to translate and to make up texts relating to business.

For students who desire to make a tourist trip to Russia the programme includes development of oral daily speech as well as presentation of the country description.

Outstanding Achievements of the Academic Staff to Date.
There are three academicians, seven corresponding members of academics, 65 doctors of science, 603 candidates of science at the University.

Special Departmental Resources and Programmes
The University contains well-equipped laboratories and computer classes; a well-stocked library; a summer sports camp; a comfortable hostel with all convenience; a youth club; a students hospital; a sport hall and a museum of minerals.

Outstanding Local Facilities and Features
With the population of more than 500,000, Irkutsk is a commercial and cultural capital of Eastern Siberia. There are the musical theatre; the drama theatre; the circus; the philharmony; the art museum and others; sport complexes; swimming-pools.

The city is located in the banks of the Angara river, 70 km far from the unique lake Baikal.

🏛	$1,000–$1,400 pa
💲	–
👥	11 000
🏢	–
%	–
💱	State grants
👤	–
📷	–
📖	650,000 volumes
💻	Yes
?	10 km far from the airprt
🛏	$7–$12
📅	1 September 1997
✈	10 km
⏱	75 hours
🚌	5,042 km

on the net   http://www.editionxii.co.uk

RUSSIA

The Krasnoyarsk State Academy of Non-Ferrous Metals and Gold

Taught Courses

BSc Degree (four years)
Automated Manufacturing Processes and Automatic Control Systems; Economics; Management; Metallurgy; Mining Engineering; Electromechanics and Electrotechnology; Advanced Materials Sciences and Technology; Geology and Minerals Prospecting.

MSc Degree and Engineering Diploma (five years)
Automated Manufacturing Processes and Automatic Control Systems; Economics; Management; Metallurgy; Mining Engineering Electrical Engineering; Electromechanics and Electroecnology; Advanced Materials Science and Technology; Geology and Minerals Prospecting.

PhD Degree (three years)
General and Regional Geology; Petrography and Volcanology; Geology, Surveying and Prospecting of Mineral Deposits; Engineering Materials; Processing and Equipment for Gold-Working of Metals; Mining Machines; Monitoring of Technical Systems; Automated Control Systems; Industrial Heat Power Engineering Fundamental Principals of Heat Engineering; Underground Mining; Metallurgy of Non-Ferrous and Rare Metals; Casting of Metals; Pressure Working of Metals; Powder Metallurgy and Composite Materials; Political Economy; Social Structure, Social Institutions and Lifestyles.

The academic year is two 20 to 21 semesters and has three faculties: Mining, Metallurgical, and Technological. The Academy is headed by the Rector and each faculty is headed by a Dean. Most young people enter the day department of our Academy, but those who work and want to get a higher education enter the evening or extra-mural department. Many experienced professors and lecturers deliver on different subjects. Our Academy occupies two large four story buildings in the most beautiful place of our town. The buildings and six hostel are very compact It comprises a sport complex with swimming pool and a bath-house, an important library, and a cafeteria. Our Academy is the only one in Russia and in the world with such a high profile scientific research centre for non-ferrous metals. Not only it is unique for its essence, but also for processing 74 components of Mendeleev system. The Academy trains engineers in 17 specialities and in nine fields of higher education.

Special Departmental Resources and Programmes
Educational Centres, Computer centre with visual display, the centre for the development and introduction of new information technologies, the "small" engineering academy, the school for young geologists. We also have two new faculties: humanitarian and further education, and a centre for staff training of higher qualifications. The Academy has six faculties; 34 chairs, 12 branches in leading enterprises and an educational industrial centre in the town of Achinsk .

The Academy offers postgraduate studentships in 21 specialities, four councils defence of a thesis on eight specialities. The Mining Faculty trains the engineers by specialities: geology and prospecting of deposits of mineral resources; technology and technics of prospecting of deposits of mineral resources; underground mining of deposits of mineral resources.

The Metallurgical Faculty trains engineers in: dressing of mineral resources, metallurgy of non-ferrous metals; political economy; economy and management of enterprises; automatization of continuous processing and manufacturing.

The Technological Faculty trains engineers in: thermal physics, automatisation and ecology of industrial furnaces; foundry of ferrous and non-ferrous metals; metal science and heat treatment; heat treatment by pressing composites and powder materials and coatings.

Outsanding Local Facilities and Features
The Foreign Affair Office; unequalled library; the medical centre "Russj"; the experimental creative unit "Jakatelj"; a sport complex; lake Shira; a marketing centre with the enterprise "Spam".

Contact
The Krasnoyarsk State Academy of Non-Ferrous Metals and Gold
Krasnoyarsk rabochy str 95
660025 Krasnoyarsk
Russia

Tel (+7) 391 2 347 882
Fax (+7) 391 2 346 311

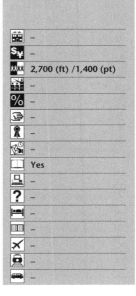

🏠	–
$¥	–
👥	2,700 (ft) /1,400 (pt)
🏢	–
%	–
✋	–
🛗	–
🕐	–
🖥	Yes
💻	–
?	–
🛏	–
▭	–
✈	–
📷	–
🚌	–

RUSSIA

Moscow State Mining University

Contact
Eugene V Kuzmin
Vice-Rector for
International Relations
Moscow State
Mining University
6 Leninskiy av.
Moscow
RF 117935
Russia

Tel (+7) 095 956 9039
Fax (+7) 095 956 9042

Taught Courses
MSc – Two years full-time, and Engineering Diploma – One year full-time
Economics of Natural Resources Use; Management; Surveying; Underground Mining; Mineral Processing; Mining and Underground Construction; Open Pit Mining; Physical Processes of Mining and Oil/Gas Production; Mining Equipment; Electrical Networks in Mining; Automation in Mining; CAD/CAM; Electrical Mining Equipment.

Research Programmes
PhD and DSc both Three years
Rock Mechanics; Mining Equipment; Underground Mining; Open Pit Mining; Mining Geology; Mine and Underground Construction; Mining Economics and Planning; Industry Management; CAD/CAM; Labour Protection and Fire Safety; Environmental Protection and Rational Use of Natural Resources.

Mining Engineering – Five years, Master's degree – Six years
Over 150 foreign students are successfully studying in the University and obtain certificates of appropriate degree of international standard.

General Information
The Moscow State Mining University was founded in June 1918 on the initiative and suggestion of a group of leading Russian mining engineers and was registered by special governmental memorandum. Its first name was "The Moscow State Mining Academy". Practically all kinds of methods related to developing the technology of the extraction of mineral resources in Russia was covered by the academic research. In 1930 the Academy created six independent Institutes: Mining, Geology, Gas and Oil, Steel, Non-Ferrous Metals and Gold, and Turf.

The Moscow Mining Institute obtained the status of University in June 1993. Over 40 000 mining engineers have been trained during its 77 years of existence.

At present the Moscow State Mining University is a leading educational and scientific organisation in Russia. One of its basic functions is the creation of teaching plans, programmes and educational literature for all the mining educational institutes in Russia.

Many scientists and professionals recognised worldwide are working in the University. Approximately 90 Doctors of Science; many professors; 470 associate professors PhD, including 340 Scientists PhD; 730 research engineers; 6 000 undergraduates; and 360 postgraduates are registered with the University. Every year over 700 undergraduates obtain the degree of mining engineering, and over 900 applicants become students.

The Moscow State Mining University prepares doctors of science and professors for leading research mining institutes in Russia. Mining engineers who have graduated at the Moscow State Mining University are working in all the regions of Russia and 66 countries in all the continents.

Over the last four years the University has started to prepare specialists with bachelor's and master's degrees in accordance with European standards. The teaching programmes have been partly converted and have become more individual, obliging, fundamental, and more flexible to the new open market relations in the mineral industry.

All courses are taught in Russian, so proficiency in the language is required: A Russian language foundation course is available. 200 foreign students are currently at the University.

Outstanding Local Facilities and Features
The University is located in the centre of Moscow, and is a beautiful historical and cultural place. Several new, eight storey buildings of the University contain more than 70 000 square metres of teaching facilities, and research laboratories are located on a 18 000 square metre area.

In 1994 the turnover was 24 billion rubles. The University has one of the most comprehensive mining libraries in Russia and a unique geological museum with a rich mineralogical collection. It publishes about 1 000 journals annually.

The University consists of six main faculties which cover the general directions of mining: mining, processing, development of underground space, the faculty of coal mining and underground construction, and the faculty of exploitation of non-ferrous and iron deposits.

Free
US $2,000
6,000
1 : 9
6%
Yes
Yes
15 August 1997
Yes
Yes
Sep. rooms by 2–3 prsns
US $600
1 Septembre 1997
40 km
–
–

North-West Polytechnical Institute

Taught Courses
- Electrosupplies
- Heating electrical stations
- Industrial heating energetics
- Electromechanics
- Electrical plants
- Industrial electronics
- Technology of mechanical engineering
- Lifting transport; Building and road-building machines and equipment
- Instrument-industry
- Automation of technologies and production
- Chemical technology of organic matters
- Chemical technology of inorganic matters
- Material science in machinebuilding
- Founding production of ferrous and non-ferrous metals
- Pressure processing of metals
- Technology of decorative processing of metals
- Equipment and technology of welding
- Automation and control of technical systems
- Computers and computing complexes; Systems and nets
- Radiotechnics
- Design and technology radioelectronical means
- Metrology; Standartisation and quality control
- Transport
- Management in transport
- Economics and management

Research Programmes
- Applied optics, including infrared and laser methods of measurements
- Semiconductor and solid state physics
- Field testing of structures
- Theoretical atom spectroscopy
- Laser technologies
- Computer aided electron circuits analysis etc

General Department Information

FACULTIES
- Faculty of Energetics
- Faculty of Mechanical Engineering
- Faculty of Chemistry and Technology of Materials
- Faculty of Informatics and Control System
- Faculty of Radioelectronics
- Faculty of Economics; Management and Traffic

Outstanding Achievements of the Academic Staff to Date
- Number of DSc – 43 (11,3%)
- Number of PhD – 263 (69,4%)
- Number of books published by academic staff during last three years: 146
- Number of articles published by academic staff during last three years: 455

Special Departmental Resourses and Programmes
NWPI together with IAESTE (International Association for Students Exchange for Technical Experience) develops the program of training for students of technical and economical specialitites. This training normaly lasts for two – three months and includes practical training in the area of students specialisation, cultural program and if necessary course of russian language.

Outstanding Local Facilities and Features
Saint-Petersburg is a big cultural and scientific centre. We have theatres, museums famous all over the world. St Petersburg occupies the 6th place in the world according to the number of tourists visiting it.

We have more than 30 Higher Educational Institutions in the city and a lot of research institutes. International life is very busy. A big number of international festivals and conferences on a wide range of topics take place in the city.

Contact
Pavel Fedorov
Chief Department of International Relations
North-West Polytechnical Institute
5 Millionnaya str
191186
Saint-Petersburg
Russia

Tel   (+7) 812 110 62 59
Fax   (+7) 812 311 60 16
Email fedorov@id.nwtu.spb.ru

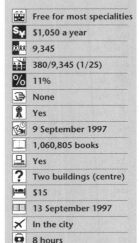

	Free for most specialities
	$1,050 a year
	9,345
	380/9,345 (1/25)
	11%
	None
	Yes
	9 September 1997
	1,060,805 books
	Yes
	Two buildings (centre)
	$15
	13 September 1997
	In the city
	8 hours
	600 km

Peoples' Friendship University of Russia
Engineering and Technology

Contact
Dmitry P Bilibin
Peoples' Friendship
University of Russia
Miklukho-Maklay St. 6
Moscow 117198
Russia

Tel   (+7) 095 434 66 41
Fax   (+7) 095 434 66 41
Email
BDP@MAIN.RUDN.RSSI.RU

Taught Courses

The list of courses taught at the MSc level includes more than 250 entries, and it can be obtained at the special request. The MSc specialisations (two years after BSc) are the following:
– Architecture and Urbanism
– Automatic Control Systems
– Computer Integrated Processes Control Systems
– Geology and Exploration of Ore and Metallic Deposits
– Hydraulic Engineering
– Industrial and Civil Construction
– Internal Combustion Engines
– Machine Building Technology
– Machine Tools and Tooling
– Manufacturing and Production
– Microprocessor Controlled Systems
– Mining Engineering
– Prospecting and Exploration of Petroleum and Gas
– Steam and Gas Turbines

Research Programmes

The following PhD programmes in Engineering and Technology (three years after MSc) are available:

ENGINEERING
– Building Constructions
– Construction Mechanics
– Hydraulics

ARCHITECTURE
– Architecture of Building and Structures

GEOLOGY AND MINERALOGY
– General and Regional Geology
– Geology Prospecting and Exploration of Oil Resources Deposits

General Departmental Information

Well-equipped laboratories and perfect computing facilities, modern methods of teaching by a highly professional academic staff. Practical classes at the best Russian industrial centres.

Outstanding Achievements of Academic Staff to Date

14 faculty members are fellows of various Russian and International academics. In 1995 the academic and research staff published 33 books and 236 articles, they received 14 patents.

Special Departmental Resources and Programmes

Excellent opportunities for carrying out the research work. 39 research projects supported by the government, and yet another 23 national and international grants received by the faculty scientists.

Outstanding Local Facilities and Features

Moscow is a beautiful city which is the Political; Financial; Scientific; Cultural and Sporting centre of Russia.

It has numerous Museums; Art galleries; Theatres and concert halls; including the world famous Kremlin; Tretyakov Gallery; Bolshoy Theatre and many others.

It also offers a wide variety of shops; cinemas; fine restaurants and cafes.

The sporting and leisure facilities of the university campus are excellent.

🏛	–
💲¥	US $1,800–US $3,800
👥	7,000
⚖	1 : 4
%	50%
🗂	Yes
🧍	Yes
🗺	1 November 1997
🖥	Yes
🖥	Yes
?	Urban
🛏	US $60–US $90
📅	15 November 1997
✈	50 km
📷	–
🚌	–

Perm State Technical University

Taught Courses
- Aircraft and Spacecraft Engineering
- Chemistry, Chemical Technology
- Civil Engineering
- Economics
- Electrical Engineering
- Environmental Protection
- Informatics, Computers and Automation
- Materials and Mechanical Engineering, Welding
- Mathematics, Mechanics
- Metallurgy
- Mining
- Philosophy and Sociology
- Theory and Methods of Teaching

Research Programmes
Functional Analysis; Functional-Differential Equations; Stressed-Deformed States and Machinery Constructions Strength; Modelling of Deformation Processes of Mono – and Polycrystals on Micro – and Mesolevels; Technological Problems of Composite; Materials Mechanics; Residual Resistance in Metals, Composite Materials and Biosystems; Mathematical Modelling, Stability and Optimal Control of Polymers Production and Treatment; Modelling of Electromagnetic Influence on the Crystallising Bars.

Mechanical Engineering; Problems of Powder and Composite Materials; Resistance to Fatigue and Cyclic Crack-Firmness; Problems of Welding; Steel and Alloys Reinforcement; Special Ways of Casting.

Automated Control Systems Projection and CAD; Modern Computer Technologies on the Certificates Base.

Technology and Use of Inorganic Sorbents; Intensification of Oil Treatment and Oil Chemistry Processes with Use of Surface-Active, Substances; Research of the Processes and Elaboration of Equipment for Crystallisation of Organic and Inorganic Substances; Elaboration of Equipment for Wet Dust-Catching; Elaboration of Equipment for Powder Materials Granulation.

Perfection of Methods of Pile Foundations Calculation According to Deformations and Carrying Capacity in Complicated Engineering-Geological Conditions with Regard to Linear and Non-linear Stage of Deforming; Problems of Environmental Protection on the Urbanised Areas; Strategy of Control of Production, Use and Neutralisation of Production and Use Wastes; Creation of Expert-Analytic System for Formation of Special Purpose Programmes on Environmental Protection.

Creation of Highly Effective Energetic Installations and Precise Equipment with the Diagnostic and Service Systems; Projection and Production of Articles of Composite Materials: Modelling of Processes, Forecasting of Features, Optimisation of Parameters; Elaboration of Resource-Protecting and Ecologically Safe High Technologies in the Sphere of Aerospace Equipment Production.

Processes of Self-Organisation and Structure-Formation in Inorganic Composite Materials Containing Crystalline and Amorphous Phases (Glass, Ceramics, Silicate Binding Substances); Town Planning: District Planning, Planning Organisation of Towns and Villages; Influence of Physical-Climatic Conditions on the State of Fencing Constructions; Ecology of Buildings: air-conditioning, Protection of Atmosphere from Industrial Wastes; Creation and Optimisation of Systems of Buildings Microclimate Provision.

Theory, Methodology and History of Sociology; Sociology of an Individual; Social Forecasting; Social Structure of Modern Society; Sociology and Philosophy of Culture; Sociology of Modernisation; Sociology of Labour; Political Sociology; Dialectics: Ontological, Logic-Geosiological and Valuation Aspects; Philosophy of Engineering Education; Theory and Methods of Foreign Languages Teaching; Economics and Enterprise Management; Religion.

Physical Processes of Mining; Perfection of Oil and Gas Deposits Exploitations; Perfection of Technologies of Underground Mining of Layer Deposits; Increase of Works Safety in Conditions of Gas-Permeability; Use of Mathematical Methods in Search and Prospecting of Oil and Gas Deposits; Interpretation of Geophysical Data in Oil and Gas Prospecting Works; Electrical Complexes and Systems,Their Control and Regulation; Automation of Technological Processes of Mining Plants; Use of Mathematical Methods and Modelling in Scientific Research; Perfection of Brake Constructions of Mine Lifting Devices; Dynamics of Mine Lifting Devices; Life Safety; Elaboration and Research of Electric Machines.

Outstanding Local Facilities and Features
Opera and Ballet Theatres; Drama, Puppet and Youth Theatres; a Philarmony; Picture Gallery; a Circus; libraries, art exhibitions; nine higher education institutions.

Contact
Valery A Moltchanov
Head of the Department of
External Relations
Perm State Technical University
29a Komsomolsky Avenue
614600 Perm
Russia

Tel   (+7) 3422 318 433
Fax   (+7) 3422 331 147
Email mv@admin.cclearn. perm.su

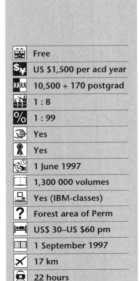

Free	
US $1,500 per acd year	
10,500 + 170 postgrad	
1 : 8	
1 : 99	
Yes	
Yes	
1 June 1997	
1,300 000 volumes	
Yes (IBM-classes)	
Forest area of Perm	
US$ 30–US $60 pm	
1 September 1997	
17 km	
22 hours	
1,400 km	

Samara State Technical University
State Committee of Russian Federation on Higher Education

Contact
Deputy Rector on
Foreign Affairs
Professor Anatoly Malyarov
Samara State Technical
University
141 Galaktionovskaia ul
443010 Samara
Russia

Tel   (+7) 8462 320 043
Fax   (+7) 8462 324 235
Email root@techun.volgacom.
samara.su
WWW
root@ie.sstu.samara.ru

Taught Courses

MSc (TWO YEARS ON TOP OF FOUR YEARS DSc)

ENGINEERING DIPLOMA (FIVE YEARS)
Applied Mathematics and Informatics; Vocational Training; Economic Management; Automated Manufacturing Processes and Automatic Control Systems; Chemical Technology and Biotechnology; Beat Power Engineering; Electrical Engineering; Electromechanics and Electrotechnology; Instrumentation Technologies; Advanced Materials Science and Technology Electrical Power Engineering; Technological Machines and Equipment; Computer Science and Engineering; Mechanical Engineering Technology Equipment and Automatic Systems; Ecology and National Use of Natural Resources; Environment Protection; Oil and Gas Technologies.

Research Programmes

PhD – THREE YEARS (AFTER MSc OR ENGINEER)

DSc – THREE YEARS (AFTER PhD)
Chemical and Molecular Physics; Physics of Solid; Chemical Physics included Physics of Burning and Explosion; Inorganic, Organic and Physical Chemistry; Chemistry and Technology of Composite Materials; Manipulators and Systems of Artificial Intelligence; Electrical Systems and Network; including their Regulation and Control; Automated Control Systems; Computer Simulation and Mathematical Modelling; Fundamental Principles of Beat Power; Engineering Drilling of Oil and Gas Wells; Metallurgy; Metal science and Treatment; Chemical Technology of Fuel and Gas; Aesthetics.

General Departmental Information
The University contains ten departments: Department of Automatics and Information Technologies; Physics; Technological; Machine-building; Engineering-economics; electromechanical; beat energetically; engineering technological; chemistry-technological; Oil-technological; Humanitarian.
   Language of consultation: Russian
   Courses taught in English: Physics; Mathematics; Fundamental Principles of Electrical Engineering; Theoretical Mechanics
   Possibilities to study Russian: Preparatory Faculty gives certificates for enrolment and study in any Russian University, Academy or Institute; fee is 1,200 USD, duration eight – ten months. There is a well-equipped workshop and excellent laboratory facilities.

Outstanding Achievements of the Academic Staff to Date
There are 93DSc, Professors, 6,500 CSc (Candidates of Sciences)ₜ Associated Professors; 25 members of Russian and international scientific academies, 8 State prize winners and laureates.
   The University publishes 20 – 25 monographs and student text-books every year. One among the latest is the book of Rector of SSTU, academician Yu Samarin "System analysis for creep in materials and structures". World Federation Publishers Comp., Atlanta, Georgia 1996, 295p.

Outstanding Local Facilities and Features
Samara is situated on the bank of the great Volgra-river: six theatres, eleven museums, Philharmonic Society with symphony orchestra, the Volga-region Chorus. Some stadiums, tennis courts and mountain-skiing trace in the city. It is good for hiking, skiing, yachting. Every summer the festival of tourist songs gatherers together thousands of fans to the countryside of Samara. Beautiful sandy beaches stretch for several kilometres. Over the Volga there are the small mountains-Giguli, it's a national park.
   Samara is connected with all parts of Russia by motorways air routes, highways and railways.

US $1,200	
US $1,800	
8,500	
1 : 8	
2	
Yes	
Yes	
20 August 1997	
Yes	
Yes	
Multi campus city	
US $5–US $10	
1 September 1997	
60 km	
19 hours	
1,000 km	

on the net   http://www.editionxii.co.uk

Tver State Technical University
Engineering

Taught Courses
BSC – four years; D Eng – five years; MSc – six years; Undergraduate course in Engineering and Medecine for foreigners – one year; Civil Engineering; Automated Manufacturing Processes and Automatic Control Systems; Mining Engineering; Chemical Technology and Biotechnology; Ground Transportation Systems; Instrumentation Technology; Electrical Power Engineering; Technology Machine and Equipment; Maintenance of Transport Facilities; Computer Sciences and Engineering; Mechanical Engineering Technology; Equipment and Automatic Systems; Environment Science; Management; Information Technology in Economics; Psychology.

Research Programmes

PhD AND DSc – THREE YEARS
Mechanics of Solids; Theoretical Physics; Physics of Solids; Thermal and Molecular Physics; Inorganic Chemistry; Chemical Kinetics and Catalysis; Tribology' Mechanical Engineering Technology; Electrical Systems and Complexes including their Control and Adjustment; Automated Control Systems; Automated Manufacturing Processes; Computer; Aided Design Systems; Computer Simulation and Mathematical Modelling in R & D; Peat Production Technology; Polymer Materials and Plastics Processing Technology; Maintenance of Transport Facilities; Footings and Foundations; Civil Engineering and Industrial Construction Technology and Management; Hydraulics and Engineering Hydrology; Labour Protection and Fire Safety; Economics.

General Departmental Information
Seven Departments at the University:
– Using of Natural Resources and Engineering Ecology
– Machine Building
– Civil and Industrial Engineering
– Automation Systems of Control
– Preliminary Department for Foreign Citizens
– Evening and Correspondence Department
– Post-graduate Training Department for improving professional skills of the specialists

Outstanding Achievement of the Academic Staff to Date
– Peat Sciences: Geological Survey of Peat Resources; Ecology of Peat – Bog Complexes; Peat Extraction and Processing; New Technologies and Products of Peat Processing
– Mechanic of Solids
– Ecology of Inner Reservoirs; Fish Protections in Hydraulic Engineering.
– Computer Simulation of Internal Processes in Metals
– Tribology
– Biotechnology
– Automated Control Systems

Special Departmental Resources and Programmes
– Super conductivity
– Transfer of Technologies
– Energetic
– Deep Processing of Coal, Oil & Gas

Outstanding Local Facilities and Features
Drama Theatre; Theatre for Youth; Puppet-play Theatre; Philharmonic Theatre; Circus; Concert Halls; Art Gallery; Museums; Cinemas; Stadium

Contact
Vice-Rector for
International Relations
Doctor Valeri L Sourinski
Tver State Technical University
Russia
170026 Tver
nab. Afanasy Nikltin 22
Russia

Tel (+7) 822 31 43 07
(+7) 822 31 15 13
Fax (+7) 822 31 43 07
Email post@politeh.tunis.tver.su

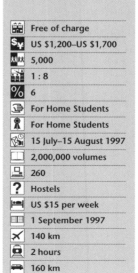

Free of charge	
US $1,200–US $1,700	
5,000	
1 : 8	
6	
For Home Students	
For Home Students	
15 July–15 August 1997	
2,000,000 volumes	
260	
Hostels	
US $15 per week	
1 September 1997	
140 km	
2 hours	
160 km	

South Africa

University of Pretoria
Department of Engineering and Technology Management

Taught Courses
The Master's in Engineering Management (MEM) Program provides in the management needs of the practising engineer (or scientist).

The Program runs over two years full-time or three years part-time. Admission requirements include an engineering (or science) degree and at least four years experience.

The program has been designed to provide the candidate with thorough education in the field of Engineering Management, while taking individual needs in to account. For this reason, the Program has been divided into two parts: a compulsory core comprising approximately 70% of the course and elective courses comprising the remaining 30%. The compusory core provides a sound, balanced foundation in Engineering Management while the elective section allows the candidate to tailor the program to his/her specific needs. In addition to course work, the candidate is required to complete a research project in the field of Engineering Management during his/her final year. The research takes place under the guidance of an advisor.

Upon successful completion of this program, a candidate may be admitted to study for the PhD degree.

The Management of Technology (MoT) Program provides engineers and scientists with in-depth knowledge and understanding of technology and its management as is applicable to a technical environment.

The honours degree is awarded after one year of full-time or two years part-time study.

Subsequent to the taught honours degree, a candidate may be admitted to study for the research-based Master's degree.

Articulation exists between the two programs.

Research Programs
– Decision Analysis
– Innovation Strategy
– Maintenance Management
– New Product and Process Development
– Production and Operations Management
– Strategic Management of Technology

General Departmental Information
The Department exists since 1987 and is part of the Engineering Faculty with approximately 2000 students. Approximately 27000 students, of whom 7000 postgraduate, are enrolled at the University of Pretoria.

Students from across the industry spectrum are enrolled and include engineers and scientists in various stages of their careers.

Approximately 25% of the students enrolled at the Department are from foreign countries.

Non-engineering management subjects that form part of the programs are offered in collaboration with the University of Pretoria Business School.

Outstanding Achievements of Academic Staff to Date
In addition to academic and research achievements, all the professors in this Department have held senior positions in Industry. This results in a practical managerial approach in educational programs.

Special Department Resources and Programs
Close ties of co-operation exist between this Department and the Institute for Technological Innovation, also in the Faculty of Engineering. The mission of the Institute is to advance technological innovation through research, interaction with industry and support of related teaching activities.

Outstanding Local Facilities and Features.
– The scenic 24 hectare campus is situated in Pretoria, the garden city capital of South Africa
– The city is of cultural, political and technological interest
– Students enjoy the excellent, sunny climate
– Sport facilities cover 76 hectares
– The campus is situated a few hours drive from the world-renowned Kruger National Game Reserve
– Theatres, parks and museums are available in the vicinity of the campus
– Johannesburg, the gold mining city and hub of commerce and industry in Southern Africa, is linked to Pretoria by a 60km six-lane highway

Contact
Prof Antonie M de Klerk
Deparment of Engineering and Technology Management
University of Pretoria
0002
South Africa

Tel   (+27) 12 420 3530
Fax   (+27) 12 43 6218
Email antonie@ee.up.ac.za

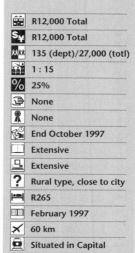

R12,000 Total
R12,000 Total
135 (dept)/27,000 (totl)
1 : 15
25%
None
None
End October 1997
Extensive
Extensive
Rural type, close to city
R265
February 1997
60 km
Situated in Capital
Situated in Capital

SOUTH AFRICA

University of Pretoria
Faculty of Engineering

Contact
Prof J A G Malherbe
Dean of the Engineering Faculty
University of Pretoria
Lynnwood Road
Pretoria
0002
South Africa

Tel   (+27) 12 420 2440
Fax   (+27) 12 43 2816
Email jagm@fanella.ee.up.ac.za
WWW
http://www.ee.up.ac.za/

Taught Courses
The programme includes an Honours' and/or Master's degree in the following specialist engineering fields: Industrial; Chemical; Electrical and Electronic; Agricultural; Mechanical; Metallurgical; Mining; Civil; Engineering and Technology Management

Research Programmes
Research programmes area are:
- Operations Research/Simulation Modelling; Logistics; Robotics; Industrial Engineering; Manufacturing
- Environmental Engineering (Water Utilisation and Air Pollution Control); Control Engineering; Polymer Science and Engineering; Corrosion
- Bio-engineering; Measurement and Control; Electromagnetics; Electronics and Micro-Electronics; Energy; Photonics; Computer and Pattern Recognition; Signals and Telecommunications
- Soil Dynamics and Farm Machinery; Processing of Agricultural Products; Irrigation and Drainage; Farm Structures
- Simulation of Fluid and Heat Transfer; Optimisation of Structures and Mechanisms; Vibration, Noise and Dynamics; Fatigue, Simulation Testing and Fracture Mechanics; Simulation of Vehicle Dynamics; Manufacturing and Manufacturing Systems
- Extractive Metallurgy; Corrosion; Welding Engineering
- Rock Mechanics; Coal Mining; Surface and Underground Mining; Mine Ventilation; Drilling and Blasting
- Transportation; Water Resources; Structures, Geotechnical and Urban Engineering
- Technology Management; Innovation Strategy; New Product Development; Maintenance Management; Project Management; Production and Operations Management; Decision Analysis

General Departmental Information
Engineering is a professional discipline and consequently the connection between research and the activities of industry is vitally important. Expertise is gained by staff and students and technology is transferred between academia and industry both on contract research and open research funded by government agencies.

The faculty has an excellent record of publications in leading international journals and the majority of academic staff are active in research and contribute to international conferences. Since the faculty 's formation in 1956, it has expanded to become the largest engineering faculty in South Africa. At present postgraduate students number approximately 700.

Limited financial aid could be made available in the form of teaching assistantships based on academic merit. Prospective students should submit a completed application form, a transcript and a $10 fee. The GRE tests are required and scores on the TOEFL must be submitted.

Outstanding achievements of the Academic Staff to date
The Faculty has a number of outstanding researchers and achievers who have received numerous awards locally and abroad.

Special Departmental Resources and Programmes
A wide range of specialised laboratories, equipment and extensive computing facilities are available. The Institute for Technological Innovation and the Laboratory for Advanced Engineering are both specialised institutes which contribute to keeping the faculty at the forefront of technology. The Department of Engineering and Technology Management is, at present, the only department of it's kind at a South African university. The Sasol Laboratory of the Department of Mechanical Engineering, Electrical and Electronic Engineering's Antenna compact Range (the only one in the Southern Hempisphere) and Civil Engineering's Concrete Laboratory are further examples of the modern, on-campus research centres which provide students with access to leading experimental and computer facilities, preparing them for professional practice and leadership.

Outstanding local Facilities and Features
The University is situated in the Capital City of South Africa, in the hub of the best entertainment and stimulation offered in the country. The State Theatre and National Arts Museum are only two examples of world-class cultural institutions in Pretoria. The city boasts sporting facilities which are on a par with the best offered in the world.

🏫	$3,500 (approximately)
💲	$3,500 (approximately)
👥	700 (postgrd) / 27,000
👤	1 : 13
%	–
🖥	Yes
👤	Yes
🗓	30 September annually
▢	Extensive
🖥	Comprehensive
?	Residential
🛏	$3,500 (per annum)
📅	End of January (annually)
✈	50 minutes by car
🚇	University in capital
🚌	University in capital

124

University of the Witwatersrand, Johannesburg
Graduate School of Engineering

Degree Programmes
The Graduate School of Engineering offers by full-time or part-time study specialised programmes in the disciplines of Aeronautical Engineering, Chemical Engineering, Civil Engineering, Electrical Engineering, Industrial Engineering, Mechanical Engineering, Metallurgy and Materials Engineering and Mining Engineering which lead to:

GRADUATE DIPLOMA IN ENGINEERING – GDE
– by advanced coursework only

MASTER OF SCIENCE IN ENGINEERING – MSc(ENG)
– by research only
– by advanced coursework only
– by advanced coursework and a project
– in Engineering Management by advanced coursework
– for holders of a three-year appropriate BSc degree

DOCTOR OF PHILOSOPHY – PhD

Fields of Research
There are approximately 40 research fields. These include Geotechnical and Materials Engineering; Structural Engineering; Water Engineering; Project Management; Construction Management and Municipal Engineering; Communications Engineering; Control Engineering; Software Engineering; Power Engineering; Aeronautical Engineering; Applied Mechanics; Applied Thermodynamics and Heat Transfer; Biomechanics; Fluid Mechanics; Mechanics of Solids; Composite Materials; Industrial Engineering; Reactor Design; Hydrometallurgy; Modelling, Optimisation and Control of Chemical Processes; Combustion; Catalysis; Biomedical Engineering; Waste water treatment; Electrochemical Engineering; Alloy Development; Coal research; Corrosion and Wear; Fatigue and Fracture Mechanics; High Temperature Processing and Pyrometallurgy; Materials Processing; Minerals Processing; Mineral Economics; Rock Mechanics; Environmental Control in Mining (Surface, Underground); Transportation, Blasting.

General Information
University of the Witwatersrand was founded in 1922, but its origins can be traced back to 1896 when the School of Mines was established. It is the largest English speaking university in South Africa and through the years has established an excellent international reputation. All the first degrees offered in the Faculty of Engineering have international accreditation and recognition. Graduate School of Engineering enjoys excellent laboratory, computing and library facilities and a conducive research and educational atmosphere. The range of soundly based programmes and research are industrially relevant and most of the projects enjoy financial support from various industries.

University Merit bursaries, Foundation for Research and Development (FRD) bursaries and stipends from industrially sponsored research projects may be available to qualified candidates.

In addition, the school provides the opportunity to study for non-degree purposes to enable practising engineers to keep abreast of the current and new technologies through the Division of Continuing Engineering Education.

Outstanding Achievements of the Academic Staff to Date
The faculty has a very good record of publications and industrially sponsored research programmes. The staff are also very active in international conferences.

Special Departmental Resources and Programmes
There are purpose-built research laboratories throughout the school. All students have access to computing facilities and networks.

Outstanding Local Facilities and Features
The university is located at the heart of Johannesburg, the Golden City. The attractive campus offers a variety of cultural, sport and outdoor activities. Johannesburg enjoys an excellent climate round the year and is a multi-cultural metropolis with a host of attractions and excellent recreational facilities. Within a 100 km radius of the city lies the world famous gold and platinum containing Reefs.

Contact
Ms Helen Glover
Postgraduate Officer
Graduate School of Engineering
University of the Witwatersrand,
Johannesburg
Private Bag 3
WITS 2050
South Africa

Tel   (+27) 11 716 5174
Fax   (+27) 11 716 5476
Email
glover@egoli.min.wits.ac.za

🏫	R3,600–R12,000 pa
💲	R3,600–R12,000 pa
👥	800 (approx)
📊	1 : 8
%	15%
📖	Yes
🎓	–
📅	31 July/31 Jan 1997
⬜	Yes
🖥	Yes
?	Urban
🛏	R250 (approx)
📅	January/July 1997
✈	25 km
☎	–
🚗	60 km

SOUTH AFRICA

University of Pretoria
Department of Engineering and Technology Management

Contact
Prof Antonie M de Klerk
Deparment of Engineering and
Technology Management
University of Pretoria
0002
South Africa

Tel   (+27) 12 420 3530
Fax   (+27) 12 43 6218
Email antonie@ee.up.ac.za

Taught Courses
The Master's in Engineering Management (MEM) Program provides in the management needs of the practising engineer (or scientist).

The Program runs over two years full-time or three years part-time. Admission requirements include an engineering (or science) degree and at least four years experience.

The program has been designed to provide the candidate with thorough education in the field of Engineering Management, while taking individual needs in to account. For this reason, the Program has been divided into two parts: a compulsory core comprising approximately 70% of the course and elective courses comprising the remaining 30%. The compusory core provides a sound, balanced foundation in Engineering Management while the elective section allows the candidate to tailor the program to his/her specific needs. In addition to course work, the candidate is required to complete a research project in the field of Engineering Management during his/her final year. The research takes place under the guidance of an advisor.

Upon successful completion of this program, a candidate may be admitted to study for the PhD degree.

The Management of Technology (MoT) Program provides engineers and scientists with in-depth knowledge and understanding of technology and its management as is applicable to a technical environment.

The honours degree is awarded after one year of full-time or two years part-time study.

Subsequent to the taught honours degree, a candidate may be admitted to study for the research-based Master's degree.

Articulation exists between the two programs.

Research Programs
- Decision Analysis
- Innovation Strategy
- Maintenance Management
- New Product and Process Development
- Production and Operations Management
- Strategic Management of Technology

General Departmental Information
The Department exists since 1987 and is part of the Engineering Faculty with approximately 2000 students. Approximately 27000 students, of whom 7000 postgraduate, are enrolled at the University of Pretoria.

Students from across the industry spectrum are enrolled and include engineers and scientists in various stages of their careers.

Approximately 25% of the students enrolled at the Department are from foreign countries.

Non-engineering management subjects that form part of the programs are offered in collaboration with the University of Pretoria Business School.

Outstanding Achievements of Academic Staff to Date
In addition to academic and research achievements, all the professors in this Department have held senior positions in Industry. This results in a practical managerial approach in educational programs.

Special Department Resources and Programs
Close ties of co-operation exist between this Department and the Institute for Technological Innovation, also in the Faculty of Engineering. The mission of the Institute is to advance technological innovation through research, interaction with industry and support of related teaching activities.

Outstanding Local Facilities and Features.
- The scenic 24 hectare campus is situated in Pretoria, the garden city capital of South Africa
- The city is of cultural, political and technological interest
- Students enjoy the excellent, sunny climate
- Sport facilities cover 76 hectares
- The campus is situated a few hours drive from the world-renowned Kruger National Game Reserve
- Theatres, parks and museums are available in the vicinity of the campus
- Johannesburg, the gold mining city and hub of commerce and industry in Southern Africa, is linked to Pretoria by a 60km six-lane highway

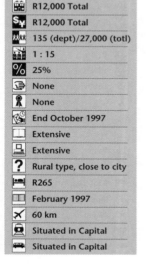

🏛	R12,000 Total
$¥	R12,000 Total
👥	135 (dept)/27,000 (totl)
👥	1 : 15
%	25%
📖	None
🧍	None
🗓	End October 1997
📖	Extensive
🖥	Extensive
?	Rural type, close to city
🛏	R265
📅	February 1997
✈	60 km
☎	Situated in Capital
🚌	Situated in Capital

Spain

Universidad Politécnica de Madrid
ETS de Ingenieros de Telecomunicación

Taught Courses

POSTGRADUATE PROGRAMME COURSES
In the following areas

- **Telematics Area**: Communication Software Design; Local and Wide Area Networks; Public Data Networks; Artificial Intelligence; Databases
- **Signal and Communication Theory Area**: Signal and Image Processing; Milimetric and Microwave Devices, Circuits and Subsystems; Radioelectrical Systems; Radar Systems
- **Radiocommunications Area**: Mobile Cellular Communications Systems; Cellular Radio Systems; GSM Systems; Satellite Communications; Digital Satellite Communications; Antenna : analysis and design

In three different levels

- **Master level:** Master in Communication Systems and Networks; January - December 1997
- **Speciality level: three – four months**: Object Oriented Software Technology; Digital; Signal Processing; Radiocommunications Systems; Communication and Switching Architecture
- **Continuing Training level**: Any course of the previous levels may be taken as a continuing training course.

Master and Speciality degrees are given by the Universidad Politécnica de Madrid.

General Information
This programme is oriented to the updating, continuing training, and recycling of university level professionals working on information technologies; technicians working in specific areas of technological development, and other people needing training in these areas. The programme is organised by the departments of Telematics Systems Engineering (DIT) and Radiocommunications, Systems and Signals (SSR).

Size of Student Body
There is a maximum of 30 students for each group at the Master level. Continuing training courses allow a typical number of five students. Speciality courses allow a maximum number of 25 students.

Apply
- Master level: November 1997; December 1997
- Specialities level: one – two weeks in advance
- Continuing training level: three days in advance

Contact
Programa de Postgrado (B-05)
ETSI Telecomunicación –
Universidad Politécnica
de Madrid
Ciudad Universitaria
s/n – 28040 – Madrid
Spain

Tel   (+34) 1 336 7364
      (+34) 1 549 5700
      ext. 414
Fax   (+34) 1 336 7363
Email
rlama@master.etsit.upm.es

🏫	990,000 ptas (Master)
💲	990,000 ptas (Master)
👥	30 (Master)
👨‍🏫	A tutor group by area
%	50%
🖨	–
👤	–
📅	Nov–Dec 97 (Master)
💻	Yes
🖥	Yes
?	University
🛏	25,000–50,000 ptas
📅	January 1997 (Master)
✈	10 km
🚇	10 mins
🚌	10 mins

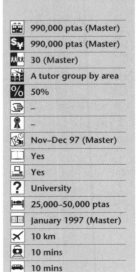

Universitat Politécnica de Catalunya
Departament d'Enginyería Electrónica

Contact

Profesor Juan Peracaula
Universitat Politècnica
de Catalunya
Departament d'Enginyerìa
Electrònica
UPC
Diagonal 647, planta 9
08028 Barcelona
Spain

Tel   (+34) 3 401 66 04
Fax   (+34) 3 401 67 56
Email peracaula@eel.upc.es

Taught Courses

PhD Programme in Electronics Engineering – Four Years

Requirements: 32 credits (at least 18 in the following courses) and PhD thesis.

Courses (credits): Analog circuits and systems (3); Analog signal processing (3); BICMOS and CMOS analog design (3); Control techniques for armonic reduction and efficiency improvement of power converters (3); Stochastic Modeling Tools for Digital Systems: Applications to Fault-Tolerant Systems (3); Digital circuits and systems (3); Electromagnetic interferences and electromagnetic compatibility (3); Heterostructure based semiconductor devices (3); Introduction to microelectronics CMOS circuits design (3); Microsystems technology (3); Modelling, simulation and control of IDC-DC converters (3); Neural networks: implementation (3); New trends in power electronics design (3); Noise and interferences in instrumentation (3); Power electronics devices and systems (3); Programmable control systems (2); Semiconductors (3); Semiconductor devices theory and technology (3); Design and High-level Synthesis of Electronic Systems (3); Testable and fault-tolerant design (3).

Research Programmes

– Instrumentation and Bioengineering
– Microelectronics Design and Digital Systems
– Power Electronics
– Semiconductor Devices

General Information

The Department of Electronics Engineering was founded in 1987 and is currently one of the largest ones at UPC. It includes more than 100 full-time faculty (seven full-time professors). The PhD programme has an annual entrance rate of about 40 students. Tuition cost is about 5,000 ptas per course. Living expenses in the area are from 75,000 to 100,000 ptas per month. Scholarships covering all expenses are available for candidates with a good curriculum from iberoamerican countries. Besides a large number of networked UNIX workstations and PCs, significant research facilities available at the department include a Semiconductor Devices Laboratory with a clean room for semiconductor device manufacturing, a solar simulator, and an instrumented point probe station, a Microelectronics laboratory including VLSI design CAD tools, a test machine and an instrumented point probe station, a Power Electronics Laboratory including high power supplies, converters and special measuring facilities, and an Instrumentation and Bioengineering Laboratory with EMC instrumentation and other advanced equipment. The department has also access to a local Supercomputing Centre equipped with Cray and IBM supercomputers and massively parallel architectures. Research is supported by industrial contracts, national research grants, and European R&D and Basic research projects.

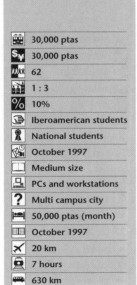

30,000 ptas	
30,000 ptas	
62	
1 : 3	
10%	
Iberoamerican students	
National students	
October 1997	
Medium size	
PCs and workstations	
Multi campus city	
50,000 ptas (month)	
October 1997	
20 km	
7 hours	
630 km	

Universitat Rovira I Virgili
Departament D' Enginyeria Química
Escola Tecnica Superior D' Enginyeria Química

UNIVERSITAT
ROVIRA I VIRGILI

Taught Courses

PhD Programme in Chemical Engineering
Requirements
- 32 credits (limit two years)
- PhD Thesis (limit five years)

Compulsory Courses (three credits each):
- Design and Analysis of Experiments
- Advanced Thermodynamics
- Transport Phenomena
- Heterogeneous Reactor Engineering
- Design and Evaluation of Research Proposals
- Seminars (one credit)

Research Areas
The rest of the courses are elective and belong to the following Research areas:
- Advanced Separation Processes
- Biochemical Engineering and Food Technology
- Computational Fluid Dynamics
- Environmental Engineering
- Fluid Mechanics and Turbulence
- Heterogeneous Catalysis and Reactivity
- Life-Cycle Analysis and Process Development
- Lignocellulosic Materials
- Polymer Science and Technology
- Engineering Thermodynamics

Qualifying Research Work (Six credits)
At the end of the second year

Required University Degrees
Chemical Engineering; Industrial Engineering; Agricultural Engineering; Chemistry; Biochemistry; Biology; Pharmacy; Physics or Mathematics.

Grants
About 10 fellowships per year to outstanding doctoral candidates are available. Please, request the necessary application forms at the contact address.

General School Information
The School of Chemical Engineering at the University Rovira i Virgili of Tarragona (URV) was created in 1995 to fulfil the need for Chemical Engineering education and research in the immediate geographical area of Tarragona that hosts one of the largest petrochemical and chemical production sites in the south of Europe. Our objective is to prepare highly educated engineers suited to successfully lead multidiscyplinari industrial research and development tasks, as well as distinguished academic careers.

The faculty belong to the Departments of Chemical Engineering and Electrical and Mechanical Engineering, and consists of 34 full-time members. Our graduate international programme has already a significant projection in its short history, with more than 30 international publications per year.

The relationship and partnership between the ETSEQ and premier international chemical companies, through common research and continuing education programmes, consulting activities, and the transfer of technologies, are a central part of our strategic quality initiative.

Contact
Doctorat
Departament D' Enginyeria Química
ETSEQ (URV)
Ctra. de Salou s/n
(Complex Educatiu)
43006 Tarragona
Spain

Tel   (+34) (9) 77 55 96 03
Fax   (+34) (9) 77 55 96 21
Email eq2@etseq.urv.es
WWW
http://www.etse.urv.es

About 147.360 ptas
About 147.360 ptas
Minimum 5
–
50%
Yes
Yes
May 97
Yes
Yes
Main City Campus
20,000–40,000 ptas
November 97
90 km
1 hour
100 km

Universidad Politécnica de Madrid
Escuela Técnica Superior de Ingenieros Industriales
Instituto Universitario de Investigación del Automóvil (INSIA)

Contact
Anselmo Morán
ETSI Industriales de Madrid
c/o José Gutiérrez Abascal, 2
28006 Madrid
Spain

Tel   (+34) 1 336 31 26
      (+34) 1 336 31 27
      (+34) 1 336 31 28
Fax   (+34) 1 336 30 03

Máster en Ingeniería de los Vehículos Automóviles
Dirigido a titulados universitarios superiores y de grado medio interesados en profundizar o actualizar sus conocimientos en las tecnologías asociadas al Sector de la Automoción.

Pone énfasis en el estudio de las tres áreas fundamentales de la actividad de las industrias automovilísticas: diseño, fabricación y comercialización tanto de vehículos como de sistemas y componentes. Se estudia el desarrollo y aplicación de las más recientes y avanzadas tecnologías en esas áreas. Se fijan unas bases sólidas para la comprensión del marco de relaciones dentro del cual se desarrolla la actividad empresarial en el sector.

Duración del Curso
Dos cursos lectivos de ocho meses cada uno, 600 horas lectivas en total. Es un programa a tiempo parcial que resulta compatible con otras actividades profesionales.
–   Horario de clases: 12 horas semanales, jueves y viernes tarde.
–   Se realizan prácticas de laboratorio, así como conferencias y visitas a instalaciones.
–   En el segundo curso, el alumno debe realizar un Proyecto de Aplicación Fin de Máster, que estará relacionado con las materias estudiadas anteriormente.

Información General
El Ministerio de Industria y Energía patrocina y financia parcialmente el Master, así como dieciseis empresas de automoción y otros organismos públicos y privados.

Las clases tanto teóricas como prácticas son impartidas por profesores universitarios de diversas Universidades del país y extranjeras y por profesionales de la industria del automóvil.

El INSIA es un instituto que realiza investigación aplicada en el sector de automoción y seguridad de tráfico, además de otra formación de posgrado en estas áreas.

Solicitudes
Antes del 10 de Octubre 1997.

ENGLISH

Master in Automobile Engineering
The course is designed for people with university degrees who are interested in extending or updating their knowledge in the technologies related to the automobile field.

The course puts an emphasis on the study of the three fundamental areas of the car industry: design, production and sales of both the cars and their systems and parts. The development and application of the most recent and advanced technologies in these areas are studied. The course will establish solid foundations for the comprehension of the framework of relationships in which business connected in this field takes place .

Length of the course
Two academic years of eight months each, 600 hours in total. It is a part-time programme that is compatible with other professional activities.
–   Timetable: 12 hours a week, Thursday and Friday afternoons.
–   Laboratory practices are organised, together with special lectures and visits to different installations.
–   During the second year, the student must undertake a Practical Project to finish the Master, which will be related to the subjects that have been studied.

General Information
The Ministry of Industry and Energy partially sponsors and finances the Master, as well as 16 automobile companies and other public and private institutions.

Classes are taught by University teachers and professors from national and foreign universities, and by professionals from the automobile industry.

The INSIA is an institute that undertakes applied research in the automotive and road safety field, as well as other postgraduate training in these areas.

Apply
Before 10th October 1997.

on the net   http://www.editionxii.co.uk

Sweden

SWEDEN

Royal Institute of Technology
(Kungl Tekniska Högskolan (KTH))
School of Architecture, Surveying and Civil Engineering

KUNGL
TEKNISKA
HÖGSKOLAN

Taught Courses

MSc IN ARCHITECTURE
The aim of architectural education states that on one hand we shall teach the art of building and architecture in order to serve the society we live in, while on the other we shall take care of and preserve our historical heritage.

MSc IN SURVEYING
Surveying consists of two parts. The first part deals with mathematics, computer science, land and civil engineering, ecology and geoinformatics. In the second part the student selects one of the following specialisations: Surveying and Mapping; Land Management; Real Estate Economics and Management; Town and Country Planning or Environmental Engineering and Sustainable Infrastructure.

MSc IN CIVIL ENGINEERING
Civil Engineering also consists of two parts. The first part deals with mathematics, mechanics, physics, chemistry, computer science/programming and basic courses of the scientific fields of civil engineering. In the second part the student selects courses within one of the following seven branches: Buildings and Indoor Climate; Civil Construction Engineering; Construction Management and Economics; Environmental Engineering and Sustainable Infrastructure; Infrastructure of Traffic; Planning and Structural Engineering.

Research Programmes

ARCHITECTURE
Research activities focus on the development of architectural criticism and theory relevant to professional practice and architectural education.

SURVEYING
Research activities in the different departments are focusing, eg. in the areas of geodetic positioning, gravity field determination and geodynamics. They also focus on legal surveying, appraisal and real estate finance and management and construction management.

CIVIL ENGINEERING
Research activities are conducted within the field of urban and regional planning, highway engineering and traffic and transport planning, wood and wood properties, building materials and indoor climate, water chemistry, geotechnology, and in dimensioning, structural design, safety, assessment and so on.

General Departmental Information
The School of Architecture, Surveying and Civil Engineering is one of five schools at KTH, The Royal Institute of Technology in Stockholm. The school has three different educational programmes, and each programme encompasses four and a half years of studies or 180 credits plus practice. Constant modernisation and updating of our curriculae has led to more interest now being devoted to the human aspects of the application of technology.

Outstanding Achievements of the Academic Staff to Date
KTH provides one third of Sweden's capacity for engineering studies and technical research at post-secondary level. KTH is the largest of Sweden's six colleges of technology. KTH cooperates extensively with various research centres and sectorial research institutes. KTH is involved in several EU research projects and is very active in the European programmes Erasmus, Comett and Tempus.

Special Departmental Resources and Programmes
Both Architecture, Surveying and Civil Engineering have programmes for exchange students ie. Nordplus, Erasmus, and bilateral agreements. At the moment one specialisation is taught in English – Environmental Engineering and Sustainable Infrastructure. Our plans for 1997/1998 are that two more specialisations, Real Estate and Spatial Planning, will be taught in English.

Outstanding Local Facilities and Features
KTH is situated in Stockholm, the capital of Sweden and home to the royal family. There are many opportunities to visit good theatres, the Opera House, numerous museums, festivals and international conferences. Each week "Stockholm Information Service" publishes a paper with listings of current activities in different theatres, museums and so on.

Contact
Gunnar Ivmark
Faculty Office of the School of Architecture, Surveying and Civil Engineering
Kungl Tekniska Högskolan
S-100 44
Stockholm
Sweden

Tel   (+46) 8 790 7943
       (+46) 8 790 7900
Email ivmark@ce.kth.se
WWW   http://www.kth.se/

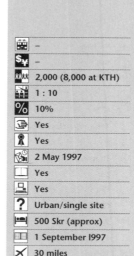

	–
	–
	2,000 (8,000 at KTH)
	1 : 10
	10%
	Yes
	Yes
	2 May 1997
	Yes
	Yes
	Urban/single site
	500 Skr (approx)
	1 September 1997
	30 miles
	Located in capital
	Located in capital

Chalmers University of Technology
School of Physics and Engineering Physics

Contact
Dr Göran Wendin
Chalmers University of
Technology
School of Physics and
Engineering Physics
S-41296 Göteborg
Sweden

Tel   (+46) 31 772 3169
Fax   (+46) 31 772 3202
Email admml@fy.chalmers.se
WWW
http://www.fy.chalmers.se

Taught Courses
The Master's programme Physics and Engineering Physics – Nanophysics (12 months full-time) is directed towards students aiming at careers in research and development – R&D – in university or industrial environments. The programme assumes that the student has a BSc degree, or equivalent, and has passed introductory courses in Electromagnetic Field Theory; Quantum Physics; Thermal Physics and Solid State Physics. The backbone of the programme is formed by five basic courses: Quantum Mechanics; Statistical Physics; Condensed Matter Physics; Computational Physics and Advanced Experimental Physics; providing a foundation for different specialisations within one of three elective programmes: Nanoelectronics and Superconductivity; Biological Physics; Advanced Materials Physics. The Physics and Engineering Physics programme is concluded with a Master's thesis conducted at R&D laboratories at Chalmers or in industry.

Research Programmes
The research programme at the School of Physics and Engineering Physics covers a fairly complete range of physics. Nanoscale physics; Superconductivity; and Advanced Materials Physics are cornerstones of basic and applied research at the School of Physics and Engineering Physics at Chalmers. The research is supported by a very strong environment of scientific instruments and technologies, including Molecular Beam Epitaxy, Scanning Tunnelling Microscopy and Atom Probe Microscopy. The school is also supported by a computer network and powerful computing facilities.

A major research facility is the Nanometer Laboratory. Artificially fabricated structures in the nanometer range will lead to a whole new world of electronics – quantum phenomena actively made use of in various kinds of electronic; optical; superconducting and molecular devices. Many future applications in telecommunications; computing; information systems; biotechnology and medicine will be based on research and development in nanoscale technologies. Knowledge of these fields will therefore be important both at university and in industry. Chalmers is now taking a major step in the direction of nanophysics and microelectronics by establishing a Centre for Microelectronics – to be operational in 1999. In this perspective, nanophysics in a wide sense, will be at the focus of the Master's Programme in Physics and Engineering Physics.

General Information
There are no tuition fees for the International Master's programme at Chalmers. There is a membership fee to the student union, which is currently 500 SEK (Swedish crowns) per year. The students have to meet their own living expenses for accommodation, food and course material (mainly books). It is estimated that a student needs about 6 800 SEK per month. No general scholarships are available at present. The school is linked with Göteborg University.

Outstanding Local Facilities and Features
Important cultural and sporting events take place in Göteborg. There are many cinemas; theatres; museums and pubs; a new opera house and a famous symphony orchestra; several parks and a big, famous amusement park. The countryside and the beautiful coast can easily be reached by the local transport system. Göteborg is a friendly and cosmopolitan city.

Detailed information about the programme, specially directed scholarships, teaching and research activities, etc, can be found from our home page at our WWW-address.

No fees	
No fees	
250	
1 : 5	
15%	
Yes – restricted eligibility	
–	
15 April 1997	
Yes	
Yes	
City campus	
2,000 SEK pm	
1 September 1997	
30 km	
4 hours	
500 km	

Ukraine

UKRAINE

State Academy of Civil Engineering and Architecture
(Dniepropetrovsk)/Sacead

Taught Courses
The Academy/SACEAD/ offers MSc and Diploma Courses in 14 specialities

– Finance and Credit
– Economics
– Production Management
– Applied Ecology (Urboecology)
– Hoisting, Construction and Road Machinery and Equipment
– Industrial and Civil Engineering
– Construction and Maintenance of Highways and Aerodromes
– Technological Process and Production Automation
– Heat and Gas Supply, Ventilation and Air Protection
– Water Supply, Sewerage, Rational Use and Protection of Water Resources
– Architecture
– Town Planning

Research Programmes

PhD and DSc (three years)
Civil Engineering – Grounds, Materials; Structures; Construction of Buildings; Town planning; Architecture; Mechanics; Metal treatment; Physics of Solids; Construction; Economics and management; Organic chemistry; Physics of Dielectrics; Work Safety.

General Department Information
The Academy is one of the largest educational establishments in Ukraine. The Academy trains about 4,000 students at the day department and 1,500 students at the correspondence department. The Academy consists of:

– Six faculties (the day department)
– Correspondence department
– Postgraduate courses for engineers and technicians
– Postgraduate courses for the instructors of technical secondary schools
– Preparatory department

In the Academy there are: 37 chairs, including 18 departments where the graduation diplomas are presented.

The Academy staff is made up 1,600 persons, including 508 instructors and 77 research workers. Students are trained by 49 professors, 240 DSc and assistant professors, candidates of Science. The value of current research contracts is over US $300,000.

Special Department Resources and Programmes
The Academy has 20 training computing rooms for 300 persons, equipped with IBM PC and IBM compatible personal computers.

The library stock exceeds 632,000 books, journals, magazines and other editions.

The Academy enjoys excellent sport facilities: a sport complex with several gymnasiums, a swimming-pool and shooting-gallery.

Contact
Prof Boris Dikarev
State Academy of Civil
Engineering and Architecture
Dniepropetrovsk / Sacead
24 A Chernishevsky Street
Dniepropetrovsk 320600
Ukraine

Tel   (+380) 562 452 372
Fax   (+380) 562 471 688

🏛	–
💲	US $1,200
👥	5,500
🏫	1 : 10
%	0,025
📖	Yes
🎓	–
📅	15 August 1997
📋	Yes
🖥	Yes
?	Urban
🛏	US $12
📚	1 September 1997
✈	15 km
🚗	11 hours
🚌	500 km

UKRAINE

Kharkov State Polytechnic University
Department of Engineering

Contact
Professor Yuri T Kostenko
Vice-Rector for
International Relations
Kharkov State Polytechnic
University
Frunze Str 21
Kharkov 310002
Ukraine

Tel   (+380) 572 432 681
Fax   (+380) 572 400 601
Email root@polytkharkov.ua
        nata@kpi.kharkov.ua
WWW
www.kpi.kharkov.ua

Taught Courses
Materials engineering; Computer facilities; Metallurgy; Power engineering; Mechanical engineering; Electrical engineering; Applied physics and mathematics; Computer science; Instruments; Electronics; Computer-aided systems; Automation in management; Chemical technologies and engineering; Management; Physical education and sports; Automation and computer-integrated engineering.

Research Programs
Applied mechanics; Physics engineering; Information science and cybernetics; Chemical engineering and technology; Science of machines; Material processing in machine-building; Power supply machine-building; Lifting transport; Electric engineering; Instruments; Information technology; Computer technology and automation; Power engineering; Metallurgy; Fat production engineering; Transport; Engineering ecology and life safety; Electronics; Economics.

Main Type of Programmes Leading to the DSc Degrees
Machines of processed by pressure, robot systems, management systems and processes; Physical chemistry engineering, processes and machinery of chemical and oil refining industries; Electrical systems, semi-conductor converters of electrical power; Physics of solids; Economics of industry and innovation; Manufacturing engineering; Drive systems, mechanical processing engineering, machine-tools and instruments; Automation of technological processes and production; Computer facilities and nets; Technology of fats, essential oils, and perfumery; Technology of non-organic substances technology; Silicate and high-melting non-metal materials; Metrology and methological control; Power machine-building; Electrical machinery and electrical engineering; Science of machines; Dynamics and strength of machines and apparatus.

General Departmental Information
Number of departments: 76, of which 53 are providing degrees: bachelor's, engineering, master's, PhD (three years following the Master's degree), Doctor's (two yeas following the PhD degree).
    The language of instruction is Russian/Ukrainian.

Outstanding Achievements of the Academic Staff to Date
300 thesis have been defended and 65 monographs and manuals published in the past five years.

Outstanding Local Facilities and Features
– Six theatres
– Three museums
– A circus
– Festivals held
– International conferences
– Gala-concerts

	134 mln krb a year
	US $1000–US $1500
	400 postgraduates
	–
	–
	Yes
	–
	In the course of a year
	–
	–
	–
	US $7.5
	–
	18 km
	5 km
	500 km

Vinnitsa State Technical University

Taught Courses

MSc – A YEAR AND A HALF (AFTER BSc)

Training in Masters of Science is provided on the following programmes Management; Mechanical Engineering; Radio Engineering; Electronics; Computer/Information Technologies; Computer Engineering; Radioelectronic Apparatus; Biomedical Systems and Apparatus; Laser and Optical Electronics Engineering; Automatics and Control Computer Systems; Electromechanical Engineering; Civil Engineering; Environmental Engineering.

Research Programmes

PhD THREE YEARS

Preparation of PhDs is provided on the following programmes; Drive Systems; Friction and Wear and Tear of Machines; Processes and Machinery of Pressure Processing; Road and Constructing Machinery; Information Measuring Systems (in science and industry); Theoretical Fundamentals of Radio engineering; Mathematical Simulation in Researches; Computers, Systems and Networks; Computer Engineering and Control Systems Elements and Devices; Industrial Thermal Power Engineering; Technology and Organisation of Industrial and Civil Engineering; Power Stations (Electrical Section), Power Engineering Systems and their Control; Networks; Economical / Mathematical Methods and Models.

General Departmental Information

All courses are taught in Russian or Ukrainian. The knowledge of one of these languages is compulsory. There is the possibility to study one of these languages (or both by options) at the foundation courses. The academic year consists of three semesters, starting in September, January and May respectively. Training in Master's degrees provided in three semesters and six months is completely devoted to the thesis preparation. The academic year of a PhD preparation is not divided into semesters. Students and postgraduate students have the possibility to publish their research results in the University scientific technical journal.

Outstanding Achievements of the Academic Staff to date

The rector, Professor B.I. Mokin, was included in the publication "Men of Achievements", 16th Edition (IBC, Cambridge) for his achievements in science in 1995. Professor A.P.Rotstein is a member of the European division of the International Society "Human Factor and Ergonomics". Professor V.A. Ogorodnikov is one of monograph authors of "Theory of forging and forming", (Moscow 1993), Professor W.Johnson from Cambridge is a co author. Professor A.A. Zhukov is a member of the Editorial Board of the journal "Cast Metals" (United Kingdom).

Special Departmental Resources and Programmes

The Department comprises schools carrying out the development of scientific and technical programmes on the following subjects noise immunity coding method; Fibonacci codes based on noise immunity measuring systems; Fourie integral method and sensitivity theory based on linear and non-linear control systems of identification, control and diagnosis; theory of fuzzy sets based man machine design and systems; complex technical system functioning, quality and efficiency evaluation criteria; computer networks synthesis image processing; Fourie fast transformation performing processors and algorithms; predetermined properties of building materials and construction production technology.

Outstanding Local Facilities and Features

Regional drama theatre; the memorial museum-estate of N.I. Pirogov (with a vault where the embalmed body of the founder of military field surgery lies); the museum of local lore; the annual festival of variety musicals "Sunsets over the Bug"; traditional international conferences at the Technical University, e.g. "Control and Management in Technical Systems", "Instruments Building" and "Humanisation and Humanitarisation of the technical education". Traditional international "Pirogov reading" are held at the Medical University and at the Museum of Pirogov. Near Vinnitsa is the centre of the hassids' pilgrimage, where Zadek Nachman, their spiritual father, is buried.

Contact
Professor Tamara B Buyalskaya
Vinnitsa State
Technical University
Khmelnitske Shosse 95
Vinnitsa 286021
Ukraine

Tel   (+380) 432 466 946
Fax   (+380) 432 465 772
Email vstu@sovam.com

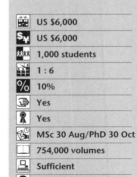

	US $6,000
	US $6,000
	1,000 students
	1 : 6
%	10%
	Yes
	Yes
	MSc 30 Aug/PhD 30 Oct
	754,000 volumes
	Sufficient
?	Compact
	US $15
	MSc 1 Sept/PhD 1 Nov
	25 km (approx)
	4 hours (approx)
	250 km

United Kingdom

UNIVERSITY
of
GLASGOW

University of Glasgow
Department of Aerospace Engineering

Taught Courses

MENG (FIVE YEARS) IN AEROSPACE ENGINEERING
Classified Honours. RAeSoc accredited to CE status

BENG (FOUR YEARS) IN AERONAUTICAL ENGINEERING
Classified Honours. RAeSoc accredited to CE status

BSC (FOUR YEARS) IN AERONAUTICAL ENGINEERING
Ordinary. RAeSoc accredited to Incorporated status
The fifth year of the MEng course involves projects in UK industry and a European academic institution. The Honours Degrees of BEng and MEng are accredited to CE status by the Institution of Mechanical Engineers.

Research Programmes
Research projects are available in the fields of: Flight dynamics, including rotorcraft; Rotorcraft aerodynamics; Computational fluid dynamics; Wind turbine aerodynamics; Space flight dynamics; Aircraft and spacecraft guidance, navigation and control.

General Departmental Information
The department is the only Aerospace Engineering department in Scotland offering research facilities and supervision for students wishing to study for the degrees of MSc or PhD by research and dissertation. The normal periods of study are two years for the MSc and three years for PhD.

Outstanding Achievements of the Academic Staff to Date
Almost all publications on dynamic re-establishment of fully attached aerofoil flow from the stalled condition are based on data derived in the department. The department's three-dimensional dynamic-stall facility (only one other comparable rig worldwide, NASA Ames), provides unrivalled data quality which is extending the department's current capability into fixed-wing agile aircraft.

The Computational Fluid Dynamics group has an international reputation for its work in the field of robust and accurate algorithms to predict aerodynamic flows modelled by the Reynolds' averaged Navier-Stokes equations. The simulations require high performance parallel computing (HPC) environments based on massively parallel processors (eg. the EPCC Cray T3D) and workstation clusters. The developed codes are crafted to the individual requirements of accuracy, fast execution and ease of use through innovative developments in high order upwind discretisation, implicit formulations with preconditioning for linear algebra solvers, and parallel algorithms.

The Rotorcraft Flight Dynamics Group is the leading UK academic centre for rotorcraft flight dynamics research. The group has developed high fidelity mathematical models, novel simulation techniques, and solutions of practical flight dynamics problems. The department's skills in this field were recognised by the Civil Aviation Authority who selected the department to study the impact of stability and control issues on light gyroplane airworthiness.

The Space Systems Group has developed innovative autonomous control methodologies for a range of spacecraft applications. A novel rendezvous and docking path planner was described by an AIAA reviewer as "a truly useful approach to handling guidance and control problems." The methodology has attracted the interest of ESA/ESTEC as an efficient means of enabling rendezvous and docking of the European ATV spacecraft at International Space Station Alpha.

The Avionics Group researches unmanned underwater vehicles supported by a group of multinational marine-technology, oil-exploration companies, including Lockheed, and the UK Government. Another Group project studies electrical coupling of flight control inceptors. This research is a collaborative project with GEC Marconi Avionics, the DRA and the DTI. The objectives of the work have major implications for the future development of avionics systems.

Special Departmental Resources and Programmes
The department's facilities include two wind tunnels and a flow visualisation tunnel; remotely piloted vehicles for in-flight experimentation and data logging; equipment for particle imaging velocimetry with data-logging and computational analysis, and extensive high speed computing facilities linked into city and nationwide computing networks.

Contact
Dr Colin McInnes
Department of Aerospace Engineering
University of Glasgow
Glasgow
G12 8QQ
United Kingdom

Tel   (+44) 141 330 5918
Fax   (+44) 141 330 5560
Email colinmc@aero.gla.ac.uk
WWW
http://www.aero.gla.ac.uk

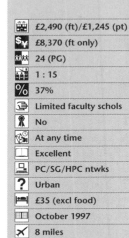

🏫	£2,490 (ft)/£1,245 (pt)
💷	£8,370 (ft only)
👥	24 (PG)
🏛	1 : 15
%	37%
📚	Limited faculty schols
🧍	No
🗓	At any time
⬜	Excellent
💻	PC/SG/HPC ntwks
?	Urban
🛏	£35 (excl food)
📅	October 1997
✈	8 miles
🚃	45 mins
🚌	44 miles (70 km)

UNITED KINGDOM

Heriot-Watt University
Edinburgh

Heriot-Watt University
Building Engineering and Surveying

Contact
Dr William A Wallace
Department of Building
Engineering and Surveying
Heriot-Watt University
Riccarton Campus
Edinburgh EH14 4AS
Scotland
United Kingdom

Tel   (+44) 131 449 5111
       ext 4647
Fax   (+44) 131 451 3161
Email
W.A.WALLACE@HW.AC.UK
WWW
http://www.hw.ac.uk.

Taught Courses

MSc BUILDING SERVICES ENGINEERING

MSc BUILDING SERVICES ENGINEERING MANAGEMENT

MSc CONSTRUCTION MANAGEMENT (CONTRACTING MANAGEMENT)

MSc CONSTRUCTION MANAGEMENT (CONTRACTS AND PROCUREMENT)

MSc CONSTRUCTION MANAGEMENT (CORPORATE STRATEGY)

MSc CONSTRUCTION MANAGEMENT (PROJECT MANAGEMENT)

MSc CONSTRUCTION MANAGEMENT (VALUE MANAGEMENT)

MSc FACILITIES MANAGEMENT AND ASSET MAINTENANCE

MSc PROPERTY ASSET MANAGEMENT

Research Programmes
– Value Management in the UK Construction Industry
– International Benchmarking in Construction
– Effects of Detergent on Drainage Water Flow, Water Flow from Roofs.
– Acoustic Performance in Classrooms
– Value Engineering
Note: All the above are EPSRC- Funded Research Contracts.

General Departmental Information
– Three Professors
– Two Industrial Professors
– 27 Academic Staff
– Eight Research Associates
– 350 Undergraduate Students
– 150 Postgraduate Students
Purpose – Built Modern Accommodation; Lecture Rooms; Laboratories; Wind Tunnel; Anecoic Chamber; State Of The Art Computing Facilities; Pleasant Modern Accommodation.
    All Postgraduate Courses are available in Full-Time, Part-Time and Distance Learning modes of study.

Outstanding Achievements of the Academic Staff to Date
Awarded Seven Research Contracts over the past six months, very high Publication and Research Contract success rate. The Graduate Programme in Construction Management has been extremely successful since it was launched in 1977. The Department is being well marketed and it is growing rapidly.
    The Postgraduate Courses are attracting growing numbers of increasingly high calibre students.

Special Departmental Resources and Programmes
The Department is the Flagship Construction Department in Scotland. The MSc in Construction Management holds a record nine EPSRC advanced course studentships. The Department has the finest acoustics equipment in Scotland including a fully equipped anecoic chamber.

Outstanding Local Facilities and Features
Edinburgh Old Town Designated European Special Heritage Status (Only one in the UK). Edinburgh is a world famous city attracting millions of visitors each year. It has unrivalled architecture and very detailed historic and cultural links. It is the ancient capital of Scotland and is famous for it's culture, artistry and creativity The Edinburgh Festival is world famous.

	£2,430 (ft)/£2,500 (dl)
	£8,400 (ft)/£2,500 (dl)
	350 Under/150 Post
	1:100
%	30%
	Yes
	Yes
	Varies
	Excellent
	Excellent
?	Suburban/Rural
	£40
	October
	3 miles
	8 minutes
	6 miles

University of Strathclyde
Bioengineering Unit

Taught Courses
Options for study in Bioengineering

PhD (THREE YEARS)

MPHIL (ONE YEAR)

MSc IN BIOENGINEERING (ONE YEAR)

INTERNATIONAL MSc IN ARTIFICIAL ORGANS (TWO YEARS)

DIPLOMA (10 MONTHS)

MSc and Diploma courses are based on a modular credit based system and include a research project. MSc students are required to attain 14 credits while Diploma students must gain 10 credits. Students select up to eight credits from a list of Conversion and Research Methodology credits including: Medical Science for engineering or non-life science graduates (four credits); Engineering Science for life science graduates (four credits); Statistics and Experimental Design (two credits); Bioengineering and Health Care (one credit). The remaining credits are obtained from a list of 15 extension credits including: Biomechanics/Rehabilitation; Biomedical Materials; Artificial Organs; Toxicology; Electronics; Instrumentation; Signal and Image Processing. PhD and MPhil candidates undertake selected classes with the approval of the Head of the Bioengineering Unit.

Research Programmes
All students complete a research project and submit a thesis. In the case of students attending the International MSc in Artificial Organs up to one year of research work related to Artificial Organs will be completed in research laboratories outwith the University of Strathclyde. The major research interests of the Unit are: Rehabilitation Engineering, including locomotion biomechanics, prosthetics and orthotics, functional electrical stimulation and motor control. Cell, Tissue and Organ Engineering covers: biomedical materials and their evaluation, artificial organ design, hybrid artificial organs, tissue mechanics, cell and tissue culture methods and applications. Instrumentation includes; equipment for clinical measurement and functional evaluation; laser techniques for the treatment of tattoos and port wine stains; electrophysiological measurements; artificial intelligence for pattern recognition.

General Departmental Information
The Bioengineering Unit is an international centre of excellence for postgraduate education and research which is directed at the application of the methods and ideas of physical and biological sciences and engineering in medicine and surgery. The postgraduate instructional course in Bioengineering is unique in its breath of coverage of topics in the field, reflecting the diversity of research ongoing in the Bioengineering Unit. The academic staff have a wide range of professional disciplines and there is extensive worldwide research collaboration within the Health Care Industry and in Clinical Medicine.

Outstanding Achievements of the Academic Staff to Date
The Bioengineering Unit received the highest grading in the last two UK wide research assessment exercises. The assessment reflects the quality and the activity of the academic staff who have extensive involvement in professional and scientific activities throughout the world.

Special Departmental Resources and Programmes
The Bioengineering Unit occupies some 3,500 square metres of floor space. Research equipment includes a fully equipped gait laboratory with two Kistler force platforms and multi-camera TV computer coordinate acquisition systems. There are a range of limb force transducers, and prosthetic/orthotic fitting facilities. There is a mechanical test laboratory containing two Instron universal test machines, scanning electron microscope with X-ray analysis facilities, laser confocal microscope and low shear blood viscometers. Artificial organ performance tests and cell culture are associated with blood gas analysers, Coulter particle counters, HPCL systems, gene pulsars, UV-VLS scanning spectrophotometer, liquid scintillator analyser and a wide range of purpose built electronic instrumentation, including Functional Electrical Stimulation equipment.

Outstanding Local Facilities and Features
The Arts Council of Great Britain awarded Glasgow (Scotland's largest city) the title of United Kingdom City of Architecture and Design. It is the home of Scottish Opera, Scottish Ballet and The Royal Scottish National Orchestra. There is a thriving and diverse arts programme in Glasgow supported by a number of popular arts and music festivals. An excellent choice of pubs, clubs and restaurants compliments abundant sports and leisure facilities (indoor and outdoor) and visitors to the city have easy access to Scotland's many tourist attractions.

Contact
Dr B A Conway
Bioengineering Unit
University of Strathclyde
Glasgow GA 0NW
Scotland
United Kingdom

Tel   (+44) 141 52 4400
       ext 3316
Fax (+44) 141 552 6098
Email b.a.conway@strath.ac.uk
WWW
http://www.strath.ac.uk/
Departments/Bioeng/

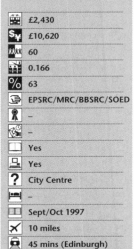

£2,430	
£10,620	
60	
0.166	
63	
EPSRC/MRC/BBSRC/SOED	
–	
–	
Yes	
Yes	
City Centre	
–	
Sept/Oct 1997	
10 miles	
45 mins (Edinburgh)	
50 miles (Edinburgh)	

Loughborough University
Department of Chemical Engineering

Contact
Miss A Temple
Department of Chemical
Engineering
Loughborough University
Ashby Road
Loughborough LE11 3TU
United Kingdom

Tel   (+44)150 922 2532
Fax   (+44) 150 922 3923
Email A.Temple@Lboro.ac.uk
WWW
http://info.lboro.ac.uk/home.html

🏫	£2,490
💲	£7,760
👥	12,000
📊	1 : 2.5 (postgraduate)
%	–
🏛	Yes
🏃	–
🗓	30 June 1997 (EPSRC)
🛏	Yes
💻	Yes
?	See above
🛏	£2,185 for 50 weeks
🏢	October 1997
✈	6 miles
🚉	1.5 hours
🚌	110 miles

Taught Courses

MSc IN PROCESS CHEMICAL ENGINEERING
Twelve month modular programme commencing in early October each year. Students choose four taught modules from list below and the programme is completed with a nine month research project.
Modules available:
–   Semester one: Applied Thermodynamics; Biochemical Engineering; Bioprocessing; Air Pollution Control; Water Pollution Control; Plant Engineering; Process Control; Process Dynamics; Process Economics and System Design; Particle Technology; Separation Processes; Knowledge-based Systems
–   Semester two: Reaction Engineering, Transfer Processes

Research Programmes

MPHIL AND PHD PROGRAMMES IN:
Adsorption and ion exchange (Professor Streat); Emulsion Polymerisation (Professor Brooks); Solid Fluid Systems (Professor Buffham & Dr Mason); Expert Systems (Dr Chung); Food and Bioprocessing (Drs Hall and Shama); Aerosol Technology and Air Filtration (Dr Stenhouse); Solid Liquid Separations (Professor Wakeman, Drs Holdich, Ward and Tarleton); Membrane Processes (Drs Cumming and Holdich, Professor Wakeman); Safety and Loss Prevention (Professor Lees and Dr Rushton); Process Control (Drs Rossiter and Drott); Computation Fluid Dynamics (Dr Nassehi); Process Economics and Design (Dr Edwards).

General Departmental Information
The Department has 20 academic staff, 275 undergraduates and 50 postgraduates. It occupies 3,500 square metres which includes two high pilot plant areas, bench scale laboratories and workshop facilities. There is a large networked computer system with extensive PC facilities. Research is organised in loose groupings which reflect common staff interests. A feature is the Carbon Research group (Professor Patrick), a specialist self-funding group working in coal and carbon.

Outstanding Achievements of the Academic Staff to Date
–   Professor R J Wakeman – Fellowship of the Royal Academy of Engineering
–   Professor M Streat – Ion Exchange Award, Society of Chemical Industry Separation Science and Technology Group
–   Dr P W H Chung – Brennan Medal IChemE
–   Professor R J Wakeman – Senior Moulton Medal IChemE

Special Departmental Resources and Programmes
–   Excellent facilities for computer-based research on plant design; safety and loss prevention and fluid dynamics
–   Extensive facilities for experimental investigation in particle technology, separation processes and solid fluid systems
–   Access to the specialised measuring techniques such as electron microscopy and other complex analytical methods

Outstanding Local Facilities and Features
–   Attractive, 216 acre single-site campus
–   Outstanding sports facilities
–   Large Student Union with its own building
–   High standard good value accommodation on-campus in both self catering and fully catering halls
–   Loughborough is a small market town set in attractive countryside
–   Busy cities of Nottingham, Leicester and Derby are nearby
–   Loughborough is on a major main rail line with a frequent service to London and is close to the M1 motorway

Nottingham Trent University
Civil and Structural Engineering

Taught Courses
Part Time, Modular MSc and Diploma

MSc In Construction Engineering Design and Management
Focusing on structural aspects of building and ground engineering alongside Management topics. For Graduates in Civil Engineering, Building and related subjects to increase their effectiveness in the workplace.

MSc European Traffic and Transportation
Comprises three MSc options taught jointly by European Institutions from Holland, Norway, France, Hungary and Germany. The language of instruction is English and the courses are flexible modular programmes.

MSc Contaminated Land Management
The course will provide graduates with the range and depth of technical and management skills necessary to investigate, assess and remediate contaminated land to return brownfield sites to profitable use.

Research Programmes
- Acoustic emission techniques in seepage detection
- Construction of steepened slopes
- Formwork design, polyurethane foam in cavity walls
- Ground related disasters and failure
- Hydrobrake testing, subsidence of structures
- River hydrology and hydropower
- Shear box for geotextile fractions testing
- Tree uprooting
- Use of vegetation to stabilise slopes
- Yield line analysis of slabs

General Departmental Information
The Department of Civil and Structural Engineering lies within the Faculty of Environmental Studies. It provides vocational education of Higher National Diploma, degree and post graduate levels through a range of courses and research. The Department has a very active relationship with industry both through sponsored research and consulting activities and through general contacts.

Outstanding Local Facilities
Nottingham is a lively vibrant city situated at the heart of the country. It has a population of around 30 000 people and over 1 in 20 is a full-time student. The city has all the facilities and entertainment a student needs - excellent shops, exciting nightlife, sports facilities and a thriving arts and cultural scene.
   The University is located in the heart of the city. If your want a break from city life, Nottingham has numerous parks, picturesque village and good walking country within easy range.

Contact
Miss Christine Wensley
MSc Secretarial Assistant
Department of Civil and
Structural Engineering
The Nottingham Trent
University
Burton Street
Nottingham NG1 4BU
United Kingdom

Tel (+44) 115 948 6008
Fax (+44) 115 948 6450
Email
Chris.Wensley@FES.NTU.AC.UK

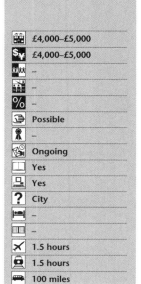

£4,000–£5,000	
£4,000–£5,000	
–	
–	
–	
Possible	
–	
Ongoing	
Yes	
Yes	
City	
–	
–	
1.5 hours	
1.5 hours	
100 miles	

UNITED KINGDOM

University
of Southampton

University of Southampton
Civil and Environmental Engineering

Contact

Mrs B Hudson
Department of Civil and
Environmental Engineering
University of Southampton
Southampton
SO17 1BJ
United Kingdom

Tel   (+44) 1703 592884
Fax   (+44) 1703 594986
Email lah@soton.ac.uk
WWW
http://www.soton.ac.uk~civilenv

💰£	£2,490 (1996/97)
💲¥£	£8,980 (1996/97)
👥	–
🏫	1 : 2
%	45–55%
📖	Yes
👤	Yes
📷	Continuous
🖥	Yes
💻	25 dept terminals
?	In landscaped valley
🛏	£35–£78 (int)/£35–£65 (ext)
🎓	MSc 1 Oct/Rsrch cont
✈	2 miles (southampton)
🚄	70 mins
🚌	70 miles

Taught Courses

The department offers four twelve month MSc courses

IRRIGATION ENGINEERING

Gives a sound understanding of irrigation within its environment and provides the necessary knowledge and skills for developing and maintaining both large and small scale irrigation schemes. It integrates the diverse disciplines needed for effective planning, design, operation and management of irrigation projects.

ENGINEERING FOR DEVELOPMENT (INFRASTRUCTURE, WATER SUPPLY AND SANITATION)

Provides training in the skills necessary for the planning and development of infrastructure, water supply and sanitation facilities for rural and peri-urban areas. It is designed for engineers working in developing countries who are concerned with the provision of integrated infrastructural facilities to meet the needs of rural communities including water supply, sanitation, waste disposal, roads, buildings, energy, and agricultural development.

PLANNING AND MANAGEMENT FOR DEVELOPMENT (AGRICULTURE OR LAND AND WATER ENGINEERING)

This course is designed: to improve the efficiency of those involved in agricultural research, extension and development enabling projects to be managed effectively; to be of value to those managing and implementing crop research; to enable new graduates to gain an appreciation of a common core of study modules. All participants are encouraged to develop an awareness of their role in the integrated functions of agricultural research, planning and management.

TRANSPORTATION, PLANNING AND ENGINEERING

Designed to give students a sound knowledge of the theory, concepts, and current practice in Transportation Planning and Engineering in both developed and developing countries. Integrating the diverse disciplines necessary for a broad appreciation of transportation, involving staff from other faculties and experts from outside organisations. It develops a range of other skills essential to the transportation profession, such an analysis and interpretation, report preparation, and communication.

Research Programmes

The department offers opportunities for postgraduate study and research in Transportation, Irrigation, Structures, Public Health, Hydraulics and Hydrology, Geotechnical Engineering, CAD, Soil Mechanics and Foundation Engineering, Coastal Engineering and Related Environmental Engineering Issues. Topics range from laboratory scale experiments to full-scale field investigations.

- *Transportation Research Group*, under the direction of Professor M McDonald, covers traffic flow and management as well as transportation modelling, planning, forecasting, project appraisal and the impact of traffic on the environment. Their work on the prediction and prevention of accidents and the personal security of travellers has been widely acclaimed.
- *Institute of Irrigation and Development Studies* works internationally on the management, operation and optimisation of irrigation systems as well as land drainage, rehabilitation and river and canal hydraulics. Problems of soil salinity and conversation are now of major concern worldwide and IIDS is an acknowledged centre of expertise in these areas.
- *Geotechnical Engineering Group* is led by Professor W Powrie, whose work ranges from computer modelling of particulate systems to applications of engineering geology. Particular areas of activity include the influence of geological structures and natural features such as fissures on geotechnical engineering designs, the geotechnics of landfills and pollution control barriers, and the behaviour of geotechnical structures such as slopes, foundations, and retaining walls. There is a close collaboration in many projects with the civil engineering industry and other universities.
- *Structures Group* specialises in composite materials and structures. Current work involves fatigue studies of sandwich panels, the extreme pressure behaviour of glass fibre reinforced resin and sandwich panels, composite connections in steel and concrete (including partial shear interaction), non-ferrous reinforcement of concrete, and composite slabs and walls. The research of the group involves both experimental programmes and theoretical studies using advanced analytical techniques.

General Departmental Information

There are 24 members of academic staff and over 40 post doctoral and research support staff. Over 200 undergraduate students follow a total of six different courses in Civil Engineering and Environment Engineering with the option of a year's study in Europe. Around 30 postgraduates pursue their studies within the various research groups.

Special Departmental Resources and Programmes

The university has a modern distributed computer infrastructure based on its own network known as SIGNET. This provides good access to the rest of the world via the Internet. Locally SIGNET connects more than 400 workstations, ranging from IBM PC compatibles, through Macintoshs and SUNs, to a nest of powerful Silicon Graphics machines used for data visualisation; it also provides access to high-power computing. Well-equipped laboratories provide extensive support for all of the subject areas listed. These include laboratories for concrete, model structures, heavy structures, public health, soil mechanics, transportation, and hydraulics and irrigation. The latter is situated at our Chilworth Laboratory, close to Southampton, where there is an exceptionally wide range of activities, including outdoor flumes and irrigation channels. The Engineering Faculty's new Graphics Centre has forty computer terminals for students use and will facilitate the teaching and learning of design and drawing skills across all the engineering disciplines.

Outstanding Local Facilities and Features

Southampton is a thriving city. Two large regional theatres, the Nuffield Theatre on campus and the Mayflower in the town both show, and sometimes premier, West End productions. The latter also has a packed programme of opera, ballet, musicals, rock, and other music concerts.

Frequent recitals and concerts are given in the purpose-built Turner Sims Concert Hall on University Campus.

Premier League football, County Cricket and athletics provide sporting entertainment and there is an annual International Boat Show. Southampton is situated on the edge of the New Forest and is ideally placed for exploring the countryside.

A centre of excellence for university research and teaching

University of Birmingham
School of Electronic and Electrical Engineering

THE UNIVERSITY
OF BIRMINGHAM

The University
The University of Birmingham is in the top five for Research Council Funding. Its Engineering Faculty, comprising five engineering schools, is one of the largest in the United Kingdom. The 200 acre parkland campus provides excellent sports and social facilities for the University's 15 000 students and is just two miles from the city centre. Birmingham is a cosmopolitan city, ideally placed for UK and international travel.

The School
The School has an international reputation and a grade 4 research rating. We have 31 academic staff, around 80 research students, of which approxiately half are from overseas, and 38 MSc students. Research income is currently about £5M, made up of European Union funding, government grants, and industrial contracts.

MPHIL AND PHD
We offer the research degrees of MPhil and PhD, over about 18 months and three years repectively, in the topics below:

- ACOUSTICS AND SONAR

- COMMUNICATIONS

- CONTROL ENGINEERING

- DIGITAL SYSTEMS AND VISION PROCESSING

- ELECTRONIC MATERIALS AND DEVICES

- POWER ENGINEERING

MSC (ENG) COURSES
Our MSc courses are one year and include teaching sessions, laboratory work, assignments and a major project.
We have three courses:

- MSC IN POWER ELECTRONICS AND DRIVES

- MSC IN ELECTRONICS AND INFORMATION TECHNOLOGY

- MSC IN COMMUNICATIONS ENGINEERING

Entry Requirements and Funding
For research degrees a minimum of an upper second class honours degree in a relevant subject is required. For taught courses a minimum of a second class honours degree in a relevant subject is required. A limited number of Research Council studentships and industrial sponsorships are available.

Facilities
Research facilities include a high precision long wave (900nm) thermal camera, a clean room for fabricating hybrid and thick film devices, a 10kW induction motor test rig for developing new AC drive strategies, a microwave range, a large network of Sun workstations, and superconductor and laser ablation facilities.

Outstanding Local Facilities and Features
World renowed Convention Centre and Symphony Hall, three leading regional theatres. The University has sports facilities amongst the best of any university in the country.

Contact
For further information please write, quoting reference PS, to:

Dr Paul Webb
Postgraduate Admissions Tutor
University of Birmingham
School of Electronic and
Electrical Engineering
Pritchatts Road
Edgbaston
Birmingham B15 2TT
United Kingdom

Tel   (+44) 121 414 4292
Fax   (+44) 121 414 4291

🏫	£2,430 (year)
💷	£8,240 (year)
👥	80 (Dept)/13,000 (Total)
🏢	–
%	50%
🖥	Yes
👤	Yes
📅	MSc mid-September 1997
▢	Yes
🖳	Yes
?	–
🛏	£28–£76
📅	9 September 1997
✈	10 miles
🚉	25 miles (station)
🚌	120 miles

University of Paisley
Department of Electronic Engineering and Physics

Contact
Alan Watt
Associate Head of Department
University of Paisley
Department of Electronic
Engineering and Physics
High Street
Paisley
Renfrewshire
PA1 2BE
United Kingdom

Tel   (+44) 141 848 3400
Fax   (+44) 141 848 3404

🏫 £2,490	
💲 £6,750	
👥 8,000	
🎓 1 : 15	
% –	
📖 –	
🧍 –	
🕐 –	
🏛 Yes	
🖥 Yes	
? –	
🛏 £30	
📅 October 1997	
✈ 5 miles	
🚗 1 hour	
🚌 50 miles	

Taught Courses
The department offers one-year Masters' courses each comprising a taught programme extending over two semesters and a project. A diploma may be awarded after the taught element.

MSc IN MICROELECTRONIC SCIENCE
This multi-disciplinary postgraduate course provides training in the theory and practice of microelectronic component manufacture. It has been designed specifically to meet the needs of those seeking a career in the semiconductor and related industries.
    For entry to the course applicants should have a first degree in electronic engineering, physics, chemistry, chemical engineering or an equivalent qualification.

MSc IN ELECTRONIC PRODUCT DESIGN FOR MANUFACTURE
This course aims to develop in graduate engineers the knowledge and skills needed to enable them to design viable electronic products. Areas as diverse as computer aided design, electromagnetic compatibility, circuit and system design and quality management are covered.
    Applicants will normally have a degree in electronic engineering or equivalent qualification.

Research Programmes
Research activities are concentrated around signal processing (SP), electromagnetic compatibility (EMC), and sensor materials and technology (SMT).

Current Projects
1. EMC Properties of "High End" Microprocessors
The research aims to obtain fundamental knowledge on the structural features at chip level which directly influence the electromagnetic (EM) properties of "high end" Digital Signal Processing (DSP) devices, in order to reduce the likelihood of expensive or dangerous failures of operation. This project benefits from established co-operation with Motorola MOS Memory and Microprocessor Division, East Kilbride, and with access to the facilities of the UKAS approved laboratory of the EMC Centre in the department.

2. EMC Properties of Microprocessors
The project, funded by Motorola Ltd, aims to identify and characterise the EMC properties of some modern microprocessor devices, and to relate the EMC effects to chip design parameters. The information obtained will be presented in a form suitable for designers to incorporate as design rules to control the EMC properties of future devices.

3. Digital Signal Processing to Enhance Speech from Noisy Reverberant Surroundings
As a result of recent work in DSP at University of Paisley (UoP) the EPSRC is supporting a project assessing a particular DSP method for speech enhancement with possible application in automobile and other environments. The underlying purpose is to improve the accuracy of human and machine speech recognition, and the quality for human listeners, where hands-free/headset-free speech input is required for everyday noisy reverberant environments. Possible beneficiaries from reliable hands-free speech-input devices are: manufacturers and users of mobile communications, physically handicapped persons, military and civil forces.

4. Digital Signal Processing for Hearing Aids
As a result of work at UoP is it believed that certain DSP techniques can be incorporated into a scheme to achieve noise and distortion reduction that may improve speech intelligibility for those suffering from a form of deafness known as sensorineural impairment.

5. Digital Hearing Aid Demonstrator
The Hearing Research Trust are funding a Postgraduate Research Studentship at the UoP providing three years support on a project to develop a digital hearing aid demonstrator for the hearing research community. The work will require the design of a prototype portable real-time digital signal processing computer system and the development and testing of selected processing algorithms for application to future hearing aids.

6. New Sensor Technologies
Investigation into the design and development of thin film pellistors for gas sensing applications.

General Departmental Information
The department has 26 academic staff and a similar number of support staff. It has a range of well-equipped laboratories designed to support the taught courses and research programmes. The department has close links with industry and is involved in a number of collaborative programmes.

Special Departmental Resources and Programmes
The department has a range of advanced specialist facilities including a clean room, scanning electron microscope with X-ray micro-analysis, X-ray diffractometer, thin film deposition equipment, CAD suites, EMC laboratory with screened room and GTEM cell.
    The department places a strong emphasis on its links with industry and delivers specialist MSc programmes to local companies.

University of Strathclyde
Department of Electronic and Electrical Engineering

Taught Courses

OPTICAL ELECTRONICS
A multi discipline MSc course involving also the Department of Pure and Applied Physics and the Department of Pure and Applied Chemistry. A 12-month full-time and 36-month part-time course from October to September designed to train new graduates in the rapidly developing technology of optoelectronics. The emphasis is directed towards systems and applications so that the graduate will be suited to design and development tasks.

ELECTRICAL POWER ENGINEERING
A 12-month full-time or 24-month part-time MSc course starting September, February and offers advanced instruction in the operation, design, control and analysis of electrical machines and electrical power systems including high-voltage technology.

INFORMATION TECHNOLOGY SYSTEMS
Offered jointly by the Department of Computer Science and Electronic and Electrical Engineering, this is a full-time course intended for graduates in science and in engineering, not already qualified in either discipline, and who are interested in applying IT in design, manufacturing or in administration.

COMMUNICATIONS, CONTROL & DIGITAL SIGNAL PROCESSING
A full-time course designed to provide a comprehensive background and thorough understanding of the advanced concepts of communications, control and digital signal processing. The course consists of a balanced mixture of theory and practice of the three disciplines and offers specialisation in either of them.

COMMUNICATIONS, TECHNOLOGY & POLICY
Modern communication systems are not only technically complex, but they have to operate in an environment in which many types of traffic are transported, and in which there are different types of participants, operators, manufacturers, service providers, regulations and users. The course is intended for those with some experience of the industry, and it is offered jointly with specialists in Computer Science and Management Science. It will combine a detailed study of the philosophy and design of modern systems, and it will highlight how they effect, and are effected by, regulatory economic and policy issues.

Research Programmes
Areas include: Power Engineering, Control Optoelectronics, Signal Processing, and Communications.

General Departmental Information
The University is recognised as one of the largest and most important institutions in the field of advanced technical education and research in the UK, while the Department is the largest of its kind in Scotland. In the past year, the Department was rated Excellent in the SHEFC Teaching Quality Assessment program.

Contact
Sheila Campbell
Department of Electronic and
Electrical Engineering
University of Strathclyde
204 George Street
Glasgow G1 1XW
Scotland
United Kingdom

Tel (+44) 141 552 4400
Fax (+44) 141 552 2487
Email
s.campbell@eee.strath.ac.uk

	£2,430 (ft)/£1,215 (pt)
	£7,750 (ft)
	211
	–
%	50%
	University Grants available
	–
	May 1997
	Excellent facilities
	Large range of equipment
?	–
	£41–£45
	October 1997
	10 minutes drive
	44 miles
	–

Brunel University
Electrical & Electronic Engineering

Contact

Postgraduate Admissions Officer
Faculty of Technology
Brunel University
Uxbridge
Middlesex UB8 3PH
United Kingdom

Tel   (+44) 189 520 3366
Fax   (+44) 189 520 3205
Email
TechPGAdmissions@brunel.ac.uk

Taught Courses

MSc in Data Communication Systems (distance learning course also available)
provides the technical and managerial knowledge and skills necessary for the initiation and management of commercial data communication networks.
Modules include: Introduction to Systems Networks; Data Security; Planning, Procurement, Regulation and Services; Software Engineering; Satellite, Optical and Mobile Communication Systems; Engineering Management; Local and Wide Area Networks; Digital Signal Processing; Practical Workshops and Assignments

MSc in Digital Systems
provides the student with a comprehensive view of digital systems in theory and practice enabling students to move into many areas of digital engineering after completion.
Modules include: CAD and Systems Design; Computer and Data Networks; Formal Methods and Automata Theory; Data Security; Software Engineering & Real Time Systems; Artificial Intelligence & Pattern Recognition; Engineering Management; Parallel Computer Architectures; Digital Signal Processing; Practical Workshops and Assignments

MSc in Intelligent Systems
provides an understanding of both natural and artificial intelligent systems spanning the areas of psychology, computer science and electrical engineering.
Modules includes: Artificial Intelligence; Artificial Vision; Software Engineering; Foundations of Intelligent Systems; Engineering Management; Neural Networks; Visual Perception; Knowledge Systems; Formal Methods and Automata Theory; AI Programming Workshops

MSc in Microelectronic Systems Design
provides the opportunity, in addition to the lecture course, to gain thorough practical design experience to those wishing to enter a career in VLSI design or CAD Design for VLST.
Modules include: Analogue IC Design; VLSI Design Techniques; CAD and Systems Design; Formal Methods and Automata Theory; Engineering Management; Parallel Computer Architectures; Digital Signal Processing; Advanced Architectures; Design Workshops.

All courses are modular and can be studied full time or part time.
Normal entry requirements for all the MSc courses are an Honours degree or equivalent in a suitable subject. Candidates with relevant alternative qualifications will also be considered.

Research Programmes
Postgraduate research opportunities leading to a degree of PhD or MPhil are offered by the department's 9 major research groups in the following areas
CAD and Test Technology; Communications; Control Systems; Medical Engineering; Microelectronics Systems Design; Neural Networks and Image Processing; Parallel Computing Technology; Power Electronics; Power Systems Analysis and Control.

General Departmental Information
The Department has always prided itself on the special relationship it has with industry. State of the art commercial software packages support much of the practical assignment and project work across the range of taught courses.

Outstanding Achievements Of The Academic Staff To Date
The Department's research activity, much of which has an international reputation, attracts considerable external funding from research councils, the European Union and industry, currently running at over £4 million.
Staff in the Department have published nearly 400 publications in the last 3 years, many of which appeared in international journals and prestigious international conferences.

Special Departmental Resources and Programmes
The laboratories are well equipped with an excellent range of facilities to support both research work and the courses. There are comprehensive computing resources within the Department in addition to those offered centrally by the University. The Department is particularly fortunate in having extensive gifts of software and hardware to enable it to undertake far reaching design projects, particularly in microelectronics.

Outstanding Local Facilities and Features
Brunel University was founded in 1966 as a mainly technological institution and has since gained an international reputation in many areas. The university campus is situated on the western edge of London close to Heathrow airport, offering easy access to London itself, to the English countryside and to international travel.

Brunel University exists to provide high quality education and research of use to the community.

🏫	£2,490
$¥	£8,195
👥	132
⚖	1 : 5
%	15%
🖥	Yes
🏆	–
🕐	End of July 1997
📖	400,000 volumes
💻	Extensive network
?	Self contained
🍴	£35–£55
📅	29 September 1997
✈	6 km (Heathrow)
🚉	20 minutes
🚌	25 km

University of Edinburgh

Taught Courses

None at present – we do, however, offer an M.Sc. by research in Analogue Design which can include taught components shared with B.Eng and M.Eng. courses. This can usefully be regarded as a part-taught M.Sc.

Research Programme

Excellent facilities exist for CAD, and for the design, fabrication, and assessment of VLSI circuits.

In addition there is an extensive computer resource within the Department, linked to the campus network, with access to commercial design, simulation and modelling software, and tools for software development, analysis and evaluation.

The Department's total research income is 1.5M pa from a wide variety of sources. Project areas include, but are not restricted to

- Signal Processing and Digital Systems
- Neural Computation and Neural VLSI
- Analogue Circuits and Systems
- Integrated Image Sensors and Processors
- Energy Generation, Distribution and Trading
- Semiconductor Technology and Microelectronic Imaging
- Speech and vision processing and recognition

General Information

In the three published surveys of research performance, the Department has achieved the top (grade five) rating. The King's Buildings Science area is located three miles South of central Edinburgh, within easy walking distance of the remainder of the University.

Outstanding Achievements of the Academic Staff to Date

All staff publish research widely and the Department's output includes textbooks on VLSI design, signal processing, neural computation and management skills. Industrial consultancy is commonplace, as are close links with industry on research programmes. The Department has "spun-off" several companies, most recently Vision Group, whose primary product is an innovative miniature, film free camera and the Centre for Communications Interface Research (CCIR) is an internationally-recognised centre for speech technology and research.

Inter Departmental and Inter University collaboration is widespread as projects extend into medical, mathematical, biological and other fields.

Internationally, we have strong links with many countries, including North America, Japan, Malaysia and Korea and of course, the rest of Europe.

Several staff are Fellows variously of the IEE, the IEEE and the Royal Society of Edinburgh. Professor Owen was named as one of Scotland's most "cited" engineering academics recently.

Staff from the Neural Systems group recently won the Scottish Hydro load prediction prize (a magnum of champagne!) and funds have just been secured to establish a world-class Microelectronic Imaging and Analysis Centre with the latest tools for problem solving in this difficult and fascinating area.

Most pleasingly, the Department was awarded first prize in the 1996 Office of Science and Technology's competition for industrial collaboration, recognising our outstanding record of industrially relevant research.

Outstanding Local Facilities and Features

Edinburgh is an historic and beautiful city which hosts festivals of film, books, jazz, folk and of course the August Edinburgh International Festival.

At no time is Edinburgh without a host of cultural events and its is a major International Conference Centre.

Access to the Scottish Highlands is straightforward and well served by road and rail links.

Special Department Resources and Programmes

- SEM measurement equipment
- Excellent cleanroom facilities and measurement capabilities support the research activities with access being available to optical, e-beam and X-ray lithography
- Nearby Edinburgh Parallel Computing Centre

Contact
Mrs L Paterson
Department of Electrical
Engineering
University of Edinburgh
Mayfield Road
Edinburgh EH9 3JL
United Kingdom

Tel   (+44) 131 650 5573
Fax   (+44) 131 650 6554
Email pgadmin@ee.ed.ac.uk
WWW
http//www.ee.ed.ac.uk

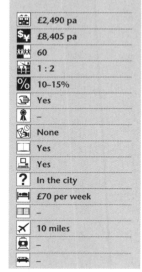

🖩	£2,490 pa
💷	£8,405 pa
👥	60
📊	1 : 2
%	10–15%
✋	Yes
🧍	–
🕐	None
📋	Yes
💻	Yes
?	In the city
🛏	£70 per week
🎞	–
✈	10 miles
🎦	–
🚌	–

UNIVERSITY OF
NEWCASTLE UPON TYNE

University of Newcastle Upon Tyne
Department of Electrical and Electronic Engineering

Contact
Prof Anthony O'Neill (E1297)
Dept of Electrical and
Electronic Engineering
Merz Court
University of Newcastle
Newcastle upon Tyne
NE1 7RU
United Kingdom

Tel   (+44) 191 222 7328
Fax   (+44) 191 222 8180
Email anthony.oneill@ncl.ac.uk
WWW
http://www.ncl.ac.uk/~neee

Taught Courses

MSc AUTOMATION AND CONTROL
Aims: To apply modern automation and control technology in industry

MSc COMMUNICATIONS AND SIGNAL PROCESSING
Aims: To provide a body of theoretical knowledge and practical experience which
is attractive to the telecommunications industries

MSc ELECTRICAL POWER
Aims: To equip graduates in electrical and electronic engineering with the
knowledge and experience to embark on a career in electrical power engineering

MSc ELECTRONICS
Aims: To provide training in electronics for science or engineering graduates
wishing to convert to a career in electronic engineering

MSc MICROELECTRONICS
Aims: To provide a bridge between a first degree and employment in the
semiconductor industry

Research Programmes
Research students study for the degrees of MPhil (two years) or PhD (three years) in
the following areas: Digital Signal Processing, Electric Drives and Machines,
Microelectronic Systems Design, Physical Electronics, Robotics and Control

General Departmental Information
The department's research activities are organised into five major research groups
reflecting the expertise of the staff. Every research student is attached to one of these
groups and is able to benefit from the supportive environment which it provides.
Each of the taught course MSc programmes is managed by one of the research
groups, so students are able to learn as part of a small special interest group.

Outstanding Achievements of the Academic Staff to Date
The quality of the department's research has been recognised by awards for papers
published in IEE Proceedings and IEEE meetings. Staff are regularly invited to serve
on the organising committees of international conferences, to give keynote
conference addresses, to serve on the editorial boards of international journals and
to give courses at international events. Professors Acarnley, Jack and Kinniment are
members of UK EPSRC Colleges.

Special Departmental Resources and Programmes
The department occupies in excess of 4,000m^2 of floor area in a purpose-built
block, which houses lecture rooms, laboratories and offices. There is a large
Student Common Room and a separate room allocated to MSc students, while
research students have personal study areas adjacent to laboratories. There are large
numbers of computer workstations and PCs with software for computer-assisted
learning, design and simulation.

Outstanding Local Facilities and Features
According to an article in *The Times* newspaper, Newcastle is among the most
popular cities in Europe with its residents. With a population of 284,000,
Newcastle serves as the commercial and cultural capital for more than two million
people and therefore has first class social, cultural, sporting and commercial
amenities. The city is the capital of one of the most naturally beautiful and
historically rich parts of the UK, and is particularly famous for its proximity to the
Hadrian's Wall World Heritage Site. The local countryside includes the hills, moors
and dales of the Northumberland National Park and the spectacular
Northumberland coast.

on the net   **http://www.editionxii.co.uk**

Leeds University

Centre for Combustion and Energy Studies

Taught Courses

MSc COMBUSTION AND ENERGY (12 MONTHS FULL-TIME)

The course provides a foundation in combustion and energy science to those who wish to pursue careers in research, development, design or plant operations involving combustion and other energy conversion processes. Candidates are required to study four compulsory taught modules, four optional modules and to conduct a research project.

First semester compulsory modules:

		Credits
CHEM 5200	Combustion and Pollution Chemistry	10
MECH 5210	Aerothermodynamics	10
FUEL 5010	Flames, Explosions and Hazards	10
COEN 5510	Computational and Experimental Methods	10
COEN 5910	Compulsory Research Project. Both programmes of study (second semester)	40

Second semester optional modules (and compulsory project):

COEN 5110	Energy Studies	10
FUEL 5020	Combustion in Boilers, Furnaces & Incinerators	10
COEN 5260	Combuston in Engines	10
MECH 5230	Computational Fluid Dynamics	10
CHEM 5210	Spectroscopy and Kinetic Modelling in Combustion	10

or up to two approved alternative modules at the discretion of the Deputy Director of the Centre for Combustion and Energy Studies. Examples include:

FUEL 5080	Control of Air Pollution	10
FUEL 3150	Fire and Dust Explosion	10
MECH 5245	Engine Tribology	10

Research Programmes

PhD research degrees are available supervised by staff from the departments of chemistry, Fuel and energy or Mechanical Enginering. Subject areas are wide ranging including: Structure of Flames; Fluid Interactions with Flames; Turbulent Combustion; Engine Combustion; Combustion Emissions; Coal Burning; Autoignition of Gases, Liquids and Solids, Flame Spread and Hazards, Laser Diagnostics, Waste Energy Conversion.

General Departmental Information

The centre is interdisciplinary, combining resources and teaching from the three constituent departments: Chemistry, Fuel and Energy and Mechanical Engineering.

Outstanding Achievements of Academic Staff to Date

There is a strong emphasis on research in the centre. Considerable progress is being made in the areas of turbulent combustion, laser diagnostics, engine combustion, coal, flame spread, hazards and autoignition. Notable books published in recent years are: *Chemical Chaos*, S K Scott, 1991 Oxford; *Combustion of Liquid Fuel Sprays*, A Williams, 1990 Butterworths.

The centre comprises the three departments:
- **Fuel Energy** – Contact Prof A Williams
  Tel: (+44) 113 233 2508
- **Chemistry** – Contact Prof M J Pilling
  Tel: (+44) 113 233 6451
- **Mechanical Engineering** – Contact Prof C G W Sheppard
  Tel: (+44) 113 233 2140

Specific enquiries within the above disciplines should be addressed to the above contact names and numbers.

Contact
Dr A C McIntosh
Deputy Director
Centre for Combustion
and Energy Studies
Fuel and Energy Department
Leeds University
Leeds LS2 9JT
United Kingdom

Tel   (+44) 113 233 2506
Fax   (+44) 113 244 0572
Email a.c.mcintosh@leeds.ac.uk

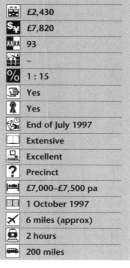

£2,430	
£7,820	
93	
–	
1 : 15	
Yes	
Yes	
End of July 1997	
Extensive	
Excellent	
Precinct	
£7,000–£7,500 pa	
1 October 1997	
6 miles (approx)	
2 hours	
200 miles	

UNITED KINGDOM

GLASGOW
CALEDONIAN
UNIVERSITY

Glasgow Caledonian University
Department of Engineering

Contact
Dr Tarsem Singh Sihria
Department of Engineering
City Campus
Cowcaddens Road
Glasgow G4 0BA
United Kingdom

Tel   (+44) 141 331 3568
Fax   (+44) 141 331 3690

Taught Courses

MSc/PGd MAINTENANCE SYSTEMS ENGINEERING AND MANAGEMENT
This is a 12-month full-time (36-month part-time) course that examines the commercial benefits gained from the implementation of maintenance technology.
   The range of topics covered in the modules constituting the taught part of the course includes:
– Condition monitoring
– Mechanical and electrical aspects of maintenance
– Reliability centred maintenance
– Risk analysis and reliability theory
– Software for maintenance
– Safety and environmental issues
– Total productive maintenance

   A major project is included, usually undertaken in an industrial environment and tailored to suit the individuals and the company's need. There are opportunities to exit at several stages of the course to obtain either a postgraduate certificate or diploma. Applicants should possess an honour's degree in a science based discipline or equivalent professional experience.

Research Programmes
Research within the Department of Engineering covers a broad spectrum of interests relevant to the needs of industry and the community at large. This can be listed into the following areas:
– Development of Condition Monitoring Instrumentation to determine the integrity of electrical plant
– Development of Mechanical Monitoring Strategies to identify on-line product quality in machining systems
– Materials and Manufacturing Systems Analysis

General Departmental Information
The department has an international reputation for research and professional development. At present it has 13 postgraduate students researching for a PhD, three postdoctoral Research Fellows following independent research programmes and approximately 70 per cent of its academic staff in the active research category as designated by the government research assessment exercise criteria.
   In addition, four students are undertaking research towards a higher degree by learning contract in industry.

Outstanding Achievements of the Academic Staff to Date
The majority of the staff are engaged in research activities and as such present papers regularly in suitable journals or at conferences on an international basis.

Special Departmental Resources and Programmes
The range of facilities/equipment available for research and development purposes reflects the broad spectrum of interests within the department. Much of the equipment is industry standard and is represented in only a few other centres around the United Kingdom.

Outstanding Local Facilities and Features
Kings Theatre; Kelvingrove Art Gallery; Burrell collections; International Jazz Festival; Botanical Gardens; Transport Museum; and the Kelvin Hall.

💷	£8,000
💲	£6,000
👥	13 (Dpt)/11,500 (Total)
	–
%	–
📖	Yes
👤	Yes
	–
📅	1 September 1997
💻	–
?	–
🛏	£40–£60
	23 September 1997
✈	10 miles
	400m (station)
	City centre

University of Hull
Computer Science

THE
UNIVERSITY
OF HULL

Taught Courses

MSc in Computer Graphics and Virtual Environments

If the world of Virtual Reality grabs your attention, this 12-month course could be for you. As a graduate in engineering or technology you will gain the further skills and experience to apply up-to-the minute VE techniques in your future engineering career.

"A unique opportunity for graduates to prepare themselves for the technological future" – JA Vince, *Virtual Reality Society*

Modules covering the core areas of Virtual Environments; Visualisation; Graphics Application Systems and Graphics Algorithms, with supporting modules in Object-Orientated Software Skills. Modules are taught and examined over two semesters, and the course is completed by individual project work leading to the dissertation. Students are encouraged to propose their own dissertation project topic, or may choose from a departmental list.

Research Programmes
The department offers research opportunities in the following priority areas:-

Graphics And Visualisation Technologies
- Computer Graphics Algorithms
- Virtual Environments
- Digital Cartography, GIS And Terrain Visualisation
- Chemical Information Systems

Software Engineering Technologies
- Software Engineering
- Parallel Computing
- Database And Multimedia

Medical Computing Systems
- Medical Information Engineering and Medical Imaging
- Computer Assisted Orthopaedic Systems

Artificial Intelligence and Systems Engineering
- End User Application
- AI Applications

Outstanding Achievements of Academic Staff to Date
The department has an established research record in many aspects of computer graphics, and a growing reputation for work in the expanding areas of Virtual Environments and Computer-Aided Orthopaedics. The latter has won the British Computer Society Award for innovation and quality in Information Technology.

Research papers are published in a wide range of journals including *Computer Aided Design*; *Computer Graphics Forum*; *BCS Computer Journal*; *Cartographic Journal*; *Journal of Molecular Graphics*; *Journal of Chemical Information and Computer Sciences*; *Virtual Reality Research, Development and Applications*; *Software Practice and Experience*; *Software Engineering Journal*; *ACM Software Engineering Notes*; *Concurrency Practice and Experience*; *Parallel Computing*; *IEE Transactions on Robotics and Control*; *Physiological Measurement*.

Departmental Information
The Department of Computer Science is of medium size, and prides itself on excellent staff-student relations. We are small enough to maintain a friendly but purposeful atmosphere, yet large enough to support a strong and well-equipped research effort. Postgraduate research is conducted within the framework of the department's research priority groups, in a co-operative atmosphere of discussion, exchange of ideas and mutual support. Each postgraduate student works under the supervision of a lead investigator throughout their study period, and progress is regularly monitored.

Research support includes the university-integrated postgraduate training scheme, and the department's research seminar series. Students are required to make formal presentations of their work in their second and third years via a regular programme of colloquia. Our departmental postgraduate forum brings postgraduate students together on a regular basis and provides further informal support and opportunities for personal development.

Special Departmental Resources and Programmes
Students benefit from dedicated working accommodation for both instructional and self-study periods, with extensive access to computing facilities. Our postgraduate work is fully supported by high specification networked computer graphics workstations, including Silicon Graphics and Sun equipment, with a wide range of applications and support software, Two general purpose computer laboratories include a large number of modern PCs with colour displays, attached to fileservers running Novell Netware. All these resources are linked by the campus network, which also gives access to further central facilities, the Library catalogue and the full spectrum of national and international nets.

Entry Requirements
Applicants should have or expect to obtain at least second class honours or equivalent in a computing or allied discipline relevant to the course or research interest.

Funding
Studentships and bursaries are available to suitably qualified applicants. Details on request.

Outstanding Local Facilities and Features
Hull is the regional centre with a wide range of excellent leisure and sporting facilities, including three theatres, two multi-screen cinemas, ice arena, swimming, athletics and team sports, a variety of art galleries and 'themed' museums.

Promoting Excellence in Education and Research

Contact
Postgraduate Admissions
(ex 97)
Department of Computer Science
University of Hull
Hull HU6 7RX
United Kingdom

Tel   (+44) 1482 465067
Fax   (+44) 1482 46666
Email pgradm@dcs.hull.ac.uk
WWW
http://www.enc.hull.ac.uk/CS/

£2,430	
£7,770	
N/A	
1 : 3	
25 : 75	
Yes	
N/A	
Any time	
Yes	
Yes	
City fringe	
£30–£35	
Sept (MSc)/Resch any month	
20 miles	
220 miles	
2.5–3 hours	

Leicester University

University of Leicester
Department of Engineering

Contact
Dr Sarah Spurgeon (ref ED12)
Postgraduate Tutor
Department of Engineering
University of Leicester
University Road
Leicester
LE1 7RH
United Kingdom

Tel   (+44) 116 252 2531
Fax   (+44) 116 252 2619
Email eon@le.ac.uk
WWW
http://www.engg.le.ac.uk

Taught Courses

MSc in Systems Engineering
This one-year course provides coverage of the advanced tools and techniques required for the modelling, analysis, control and design of complex systems which are made from interacting subsystems and yet are required to meet overall performance objectives. Graduates of the course will
– understand the many facets of Systems Engineering
– be able to analyse and design instrumentation systems, software systems, microprocessor systems and control systems
– be skilled in the use of state of the art software for control system design
– be able to integrate systems and assess their reliability
– be skilled in techniques of project management
– be practised in basic methods of research

Research Programmes
MPhil/PhD study is available in the areas of

Bioengineering, Transducers and Signal Processing
Transducers and intelligent signal processing applied to medical diagnosis and general instrumentation

Control Systems
New methods for advanced control system design – theory and applications

Electrical Power Engineering
Power electronics and wind energy recovery; breakdown in high-voltage polymeric insulating systems; electrostatic precipitators

Mechanics of Materials
Failure and deformation processes in materials; the effect upon component strength of materials processing; design and analysis of high temperature structures

Radio Systems
Long-range radio communications systems, signal propagation and direction finding

Thermofluids and Environmental Engineering
Experimental and computational fluid dynamics; combustion; water and wastewater treatment by appropriate technology.

General Departmental Information
The department has 29 academic staff including five Professors supported by seven academic related staff, approximately 20 research staff and 30 technical and clerical staff. It has a strong research tradition with in excess of 150 research papers being published in a typical year by members of the department. The department attracts considerable industrial, government and charity funding.

Special Departmental Resources and Programmes
Research activities are concentrated into six groups, each of which comprises of several academic staff, research staff and postgraduate students and provides a supportive and productive working environment. All postgraduate students belong to such a group. A number of seminar programmes are run within this structure. Close links are maintained with industry and the relevant professional institutions.

Outstanding Local Facilities and Features
Leicester is situated in the centre of England with excellent communications via road, rail and air networks. This ethnically diverse city is a cosmopolitan and thriving community. Restaurants offer cuisine from all over the world and a whole range of festivals are celebrated from the Caribbean Carnival to Diwali. Leicester is Britain's first environment city, flourishing with green initiatives. A vast range of entertainment including Western and Asian cinema, music, dance and theatre is available. Sporting attractions include premiership football, division one rugby and county championship cricket as well as basketball, horse racing, roller skating, archery and swimming. Leicester is proud to house a famous covered market, two state of the art shopping complexes and the 'Golden Mile' which is the thriving shopping street of the city's Gujarati community.

🏛	£2,490 (1996/97)
$	£8,040 (1996/97)
👥	8,500
👥	1 : 2 (PG)
%	17%
☞	Yes
🛉	–
📅	MSc Mid September
🏫	Extensive
🖥	Extensive
?	Urban
🛏	£33.81 (1996/97)
🎓	29 September 1997
✈	20 miles
🚗	70 mins
🚌	100 miles

University of Strathclyde
Faculty of Engineering

Taught Courses
There are 40 postgraduate instructional courses within the Faculty. These include: Electrical Power Engineering; Mechanics of Materials; Thermodynamics, and Fluid Mechanics; Computer-Integrated Manufacture; Bioengineering; Hydraulics, Hydrology and Coastal Dynamics; Structural Engineering; Optical Electronics; Water Engineering; Environmental Health; Computer-Aided Engineering Design; Marine Technology; Communications, Control, and Digital Signal Processing.

Research Programmes
There are approximately 30 research programme areas leading to MPhil, PhD and MArch qualifications. These include:
- Architectural and Building Aids Computer Unit
- Bioengineering
- Chemical Engineering
- Communications
- Dynamics and Control
- Energy Simulation Research
- Hydraulics and Coastal Engineering
- Industrial Control
- Manufacture and Engineering Management
- Optoelectronics
- Power Engineering
- Safety and Environmental Management
- Signal Processing
- Urban Design
- Water and Environmental Management

General Departmental Information
The Faculty is one of the largest in the United Kingdom and has the largest postgraduate population in Scotland. It comprises nine departments: *Architecture and Building Science, Bioengineering, Chemical Engineering, Civil Engineering, *Design, Manufacture and Engineering Management, *Electronic and Electrical Engineering, *Mechanical Engineering, National Centre for Prosthetics and Orthotics, Ship and Marine Technology.
*Rated 'Excellent' for teaching quality

Outstanding Achievements of the Academic Staff to Date
The Faculty has an outstanding record of publications and industrial research. This is reflected in the results of national surveys of research and in the Faculty's long-standing links with industry.

Special Departmental Resources and Programmes
There are purpose-built research laboratories throughout the Faculty. All students have access to networked computing facilities with the latest software and graphics packages. Among the specialist centres of the Faculty are: the CAD centre; Industrial Control Centre; Centre for Electrical Power Engineering; Optoelectronics Division; Safety and Environmental Management Unit; Bioengineering Unit; Architecture and Building Aids Computer Unit; Energy Systems and Mechanics of Materials. The Faculty has research contracts in excess of £14 million.

Outstanding Local Facilities and Features
Located in the heart of the city, Strathclyde is ideally placed for access to the many theatres, museums and art galleries of Glasgow. Home to the Scottish Ballet and to the Scottish Opera, Glasgow offers a wealth of cultural opportunities and attracts major international performers to its Scottish Exhibition and Conference Centre.

Contact
Susan Bridgeford
Faculty Office (Engineering)
University of Strathclyde
16 Richmond Street
Glasgow G1 1XQ
United Kingdom

Tel   (+44) 141 553 4112
Fax   (+44) 141 552 0775
Email
s.bridgeford@mis.strath.ac.uk

£2,450 (approx)	
£7,750 (approx)	
–	
–	
10%	
Yes	
–	
Continuous	
Yes	
Yes	
City centre	
£30–£40	
Flexible	
10 km	
45 mins	
40 km	

Imperial College of Science, Technology and Medicine
Centre for Environmental Technology

Contact
Assistant Registrar
(Admissions)
David Atkins
Imperial College
London
SW7 2AZ
United Kingdom

Tel   (+44) 171 594 8014
Fax   (+44) 171 594 8004
Email d.atkins@ic.ac.uk
WWW   http://www.ic.ac.uk

Taught Courses
MSc/DIPLOMA IN ENVIRONMENTAL TECHNOLOGY (ONE-YEAR FULL-TIME)
The MSc course takes students from a wide range of backgrounds and provides an in-depth training in the identification and resolution of environmental problems. It has a strong international emphasis and recruits students from both the developed and the developing world.

1) The Core Course
This is held in the first term and aims to provide a broad and interdisciplinary understanding of the natural and human environment and focuses on: The Natural Environment, Perturbed Environment, Environmental Management & Policy, Environmental Law, Economics, Statistics/Computing.

2) The Specialist Options
The following options are offered in the second term. Business and the Environment; Ecological Management; Energy Policy, Environmental Analysis and Assessment; Pollution; Water Management; Environment and Health; Global Environmental Change.

3) The Research Project
The individual research project occupies approximately 5 months and is normally based on the option selected in the second term. Projects are designed to incorporate career development and can be carried out with an external organisation. They may also be carried out abroad.

Research Programmes
In addition to the Centre's broad environmental expertise there are several groups with specialisms in environmental geochemistry, renewable natural resource management, environment and health, environmental law, policy and management, air pollution modelling and impacts, atmospheric chemistry, environmental radioactivity and radioecology, environmental decision analysis, energy policy and global environmental change. Many of ICCET's academic staff on the MSc Programme are recognised international leaders in these disciplines. Studentships are available from the ESRC, EPSRC, NERC and MAFF.

General Departmental Information
The Imperial College Centre for Environmental Technology (ICCET) was created in 1977. It was the first United Kingdom interdisciplinary academic institution for environmental technology and research. It draws upon the knowledge and experience of over eighty environmental experts in the College. One of its chief aims is to maintain an environment within which both original research on environmental protection matters and its applications to useful, practical purposes flourish. ICCET's popular Master of Science course in Environmental Technology provides environmental training for future managers, environmental engineers, scientists, policy leaders, and consultants. We also offer MPhil and PhD programmes in most areas of environmental science, engineering, economics, law and policy.

Outstanding Achievements of the Academic Staff to Date
At present, Academic staff at Imperial College have the highest per capita share of research grants in the United Kingdom. The concentration and strength of its research in science, engineering, technology and medicine give it a unique, internationally acclaimed research presence. In 1994–1995, income from research contracts and grants was £65 million.

Special Departmental Resources and Programmes
The college has large, well equipped research laboratories in all relevant departments. Students will have access to the Centre for Analytical Research in the Environment, situated at Silwood Park, an estate of 240 acres near Ascot in Berkshire. It houses an interdisciplinary research team including physicists, chemists, ecologists and aerosol and soil scientists working together on fundamental studies related to environmental issues. Students at Imperial College also have access to all of the Campus Libraries and those of the University of London, one of the most comprehensive library systems in the world.

Outstanding Local Facilities and Features
ICCET is located in South Kensington an attractive area in the centre of London. ICCET is only a few minutes from Hyde Park and Kensington Gardens and its community contains the majority of London's world class museums and its foreign embassies. As one of the worlds leading cities, London offers a vast array of amenities borne out of its exciting cosmopolitan and multicultural heritage. Major United Kingdom Government Departments, private sector firms and non-governmental organisations with an interest in environmental law and policy are only an underground train ride away. As a major global centre for art, politics, fashion, music, business, finance and academic endeavour, students will have every opportunity to take advantage of the rich cultural panorama that London has to offer.

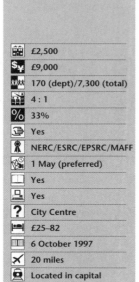

🏫	£2,500
💲	£9,000
👥	170 (dept)/7,300 (total)
👥	4 : 1
%	33%
	Yes
	NERC/ESRC/EPSRC/MAFF
🕐	1 May (preferred)
	Yes
🖥	Yes
?	City Centre
🛏	£25–82
	6 October 1997
✈	20 miles
	Located in capital
🚌	Located in capital

University of Brighton
Faculty of Engineering and Environmental Studies

Taught Courses
The University of Brighton offers a range of courses at Masters level which are both innovative and relevant to the needs of industry today. Full-time, part-time and CPD modes of study enable students to choose the course best suited to their requirements, and companies are encouraged to use them to enhance their staff development programmes.

MSc DIGITAL ELECTRONICS AND PARALLEL PROCESSING SYSTEMS
(offered jointly with the University of Sussex)
The course combines the resources of two successful institutions to provide a specialist postgraduate degree programme, focusing on real-time digital processing techniques and applications. This course is organised by the Department of Electrical and Electronic Engineering (DEEE).

MSc PRODUCT INNOVATION AND DEVELOPMENT
The course provides the skills and expertise necessary to take a wide range of products from concept through development to market. It will provide technical, business and industrial design experience that individuals and companies require for this demanding work. This course is organised by the Department of Mechanical and Manufacturing Engineering (DMME).

Research Programmes
Research at the University of Brighton is primarily applied research and focused in subject specific units which have gained national and international recognition. Strong links with industry are further enhanced by the Teaching Company Centre, one of the ten currently funded in the UK.

PhD and MPhil research opportunities are available in a number of areas including: Image Sensor Design; Neural Networks; Real-time Machine Vision Applications; Digital Audio Broadcasting; Fieldbus Protocols; Resonant Power Supply Design; High Voltage Photoconductive Switches; Automotive and Refrigeration Heat Exchangers; Computational Fluid Dynamics; Flight and Process Simulation; Vibration Monitoring; Polymer Engineering.

General Departmental Information
Both the DEEE and DMME at Brighton support their postgraduate activities via focused research units which provide a range of investigative projects to support MSc taught courses and MPhil/PhD research programmes. The departments have developed their industrial and commercial links over many years in both basic and applied research. Applied research is carried out by all of the groups, much of it being industrially funded. Funding has been received from many local and national manufacturing companies and research organisations, and collaboration extends internationally to academic and industrial partners across Europe. Industrial collaboration occurs through contract research or EPSRC/DTI teaching company schemes. The University of Brighton hosts one of the largest teaching company scheme centres in the UK with opportunities to qualify for masters degrees with minimal disruption to employment.

Outstanding Achievements of the Academic Staff to Date
Research achievements include the award of three patents, the appointment of two staff as visiting professors at other institutions, active participation of staff in professional institution (IEE and IMechE) seminar and conference organisation, and the publication of research papers in learned journals.

Special Departmental Resources and Programmes
To support its masters courses and research programmes the DEEE facilities include an ECAD/VLSI design suite, an EMC test and consultancy unit, an image processing laboratory, a full range of industry standard software and hardware development and test equipment for real-time microprocessor system design, a high voltage laboratory and partial discharge measurement systems. The department runs a Novell network supporting its PC and Sun computing suites. Special purpose laboratories in the DMME include engine test sets, climatic wind tunnel test rigs, an electron microscope facility, and particle image velocimetry (PIV) equipment for heat transfer studies. In addition the department has extensive computing facilities including an IBM PC suite, Sun work stations, and silicon graphics Indigo work stations.

Contact
Graeme Awcock, DEEE
Richard Fletcher, DMME
University of Brighton
Cockcroft Building
Brighton
BN2 4GJ
United Kingdom

Tel (+44) 1273 642208
                642300
        or 642278
Email eee@bton.ac.uk
or mecheng@bton.ac.uk
WWW
http://www.brighton.ac.uk/

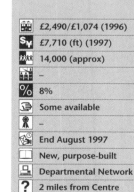

⬜	£2,490/£1,074 (1996)
⬜	£7,710 (ft) (1997)
⬜	14,000 (approx)
⬜	–
%	8%
⬜	Some available
⬜	–
⬜	End August 1997
⬜	New, purpose-built
⬜	Departmental Network
?	2 miles from Centre
⬜	From £40
⬜	September 1997
✕	25 miles (approx)
⬜	1 hour
⬜	60 miles (approx)

UNIVERSITY
OF CENTRAL
LANCASHIRE

University of Central Lancashire
Faculty of Design and Technology

Contact
Professor Norman Burrow
Faculty of Design and
Technology
University of Central Lancashire
Preston
Lancashire
PR1 2HE
United Kingdom

Tel   (+44) 1772 893160
Fax   (+44) 1772 892901
Email N G Burrow
WWW
http://www.uclan.ac.uk

🏫	£2,350
💲¥	£5,979
👥	19,693 (Uni)
👤	17.6 (Uni)
%	1% (Uni)
📖	Limited
🧍	–
🗓	September 1997
☐	Yes
💻	Yes
?	Town Centre
🛏	£35–£50
🎓	September 1997
✈	Manchester (40 miles)
⏱	3 hours
🚗	200 miles

Taught Courses
The faculty offers the following taught Masters courses
MSc Product Design
MSc Building Heritage and Conservation
MSc Electronics (Digital Signal and Image Processing)
MSc Computing

Research Programmes
The faculty welcomes applications for MPhil and PhD in the following areas
Electronics – Digital Image Processing; Digital Systems; Microelectronics and
Semiconductor Devices; Environmental Electronics: Engineering – Tribology; Product
Design; Composite Materials; Stress Analysis: Computing – Methods of Pattern
Analysis; Socio-Technical Methods; Methods for HCI; 'Soft' Systems Methodology;
Methodologies for Information Systems Analysis and Design (and for Multimedia
Design); Framework for Critical Evaluation of Methodologies; Framework for
Selection of Methodologies: Built Environment: International Procurement; Project
Management; Environmental Assessment; Construction Contract Law; Fire Safety
Engineering; Construction Materials Science; Materials and Building Performance;
Building Heritage and Conservation.

General Departmental Information
The Faculty of Design and Technology brings together the engineering departments
of Computing, Engineering and Product Design, Built Environment and Electrical
and Electronic Engineering with the design areas of art and design. It offers extensive
opportunities for postgraduate education, consultancy and research. It wishes to
encourage applied and particularly strategic research which can support the teaching
function and be of benefit to the community and industry. The faculty has Visiting
Professors from the Royal Academy of Engineering in the departments of Electrical
and Electronic Engineering and Engineering and Product Design. The faculty has
organised a number of international research conferences including a joint
conference with Beijing Institute of Technology which focused on the development
and role of women in technology, a conference on Fire Safety Engineering in Nimes
sponsored by the European Commission, and an International Product Design
conference in Preston.
    The university has a special adviser to provide information, advice and assistance
on all matters concerning international students.

Outstanding Achievements of Academic Staff to Date
Members of staff frequently present research papers at international conferences. Ian
Sherrington is a member of the IMechE Tribology Specialist Group; Nimal Jayaratna
is Chair of the British Computer Society Specialist Group on Information Systems
Methodologies, Vice President of the European Society of Projectics, Chair of the
'Soft' Systems Methodology Research Discussion Group. Barbara McManus is the
Chair of the British Computer Society North West Branch, Duncan Telfer is the Joint
Convener of the British Machine Vision Association, N W Chapter.

Special Departmental Resources and Programmes
The faculty incorporates the following centres:
–   Centre for Women in Design, Technology and Manufacture
    The aim of this centre is to encourage girls and women to positively consider a
    career in technology; to enable them to gain access to technical courses at higher
    education level and to ensure that the learning environment is appropriate for
    their needs.
–   Centre for Digital and Image Processing
–   Centre for Research in Fire and Explosion Studies
–   Laboratory for Methodology Research
It is possible to pursue the degrees of MPhil and PhD by full or part-time study.
Research students have full access to the extensive facilities and equipment available
in the various specialist laboratories in the faculty, along with full workshop and
technician support, the University Library and Computer Centre, and Internet access.
The faculty has also developed a training programme to support research students.

Coventry University
School of Engineering

Taught Courses
MSc ADVANCED MANUFACTURING SYSTEMS ENGINEERING

MSc AUTOMATION ROBOTICS AND CONTROL ENGINEERING

MSc AUTOMOTIVE ENGINEERING DESIGN AND MANUFACTURE

MSc ENGINEERING BUSINESS SYSTEMS

MSc ENGINEERING AND MANUFACTURING MANAGEMENT

MSc INFORMATION TECHNOLOGY FOR MANUFACTURE

MSc INDUSTRIAL ELECTRONICS

MSc MECHATRONICS

MSc OPERATIONAL TELECOMMUNICATIONS

MBA ENGINEERING MANAGEMENT

Research Programmes
The school has active research groups in the following areas
- Advanced joining
- Applied thermofluids
- Automotive engineering
- Cellular manufacturing
- Control theory and applications
- Integrated design
- Knowledge based engineering
- Machine vision
- Machinability
- Manufacturing and supply chain management
- Non-linear systems

General Departmental Information
The school is are of the largest centres of engineering education in the UK, with over 2,000 students, of whom 250 are on postgraduate courses, both full-time and part-time.

All courses have a strong vocational orientation and are focused on the application of engineering in practical situations. Strong links with companies such as Jaguar, Dunlop, Peugeot, Rover, GPT and Cable and Wireless ensure continuing relevance to employers' needs.

Outstanding Achievements of the Academic Staff to Date
Staff have been awarded a major role in a £29 million regional programme to enhance the competitiveness of the automotive component supply industry.

Research on automated plasma spot-welding received a Ford Technical Achievement award and is now supported by Volvo, Rover and Audi.

The school is a main partner in a 1.77 million ECU European Union project on electronics manufacturing of multichip modules.

Special Departmental Resources and Programmes
The school is well equipped with a wide variety of laboratories and computing facilities.

Postgraduate programmes operate from exclusive accommodation in a postgraduate centre comprising a suite of high quality lecture rooms; modern computing facilities; a student lounge; dedicated administrative support.

Outstanding Local Facilities and Features
Coventry is situated in the heart of England, on the edge of Shakespeare country with good road, rail and air connections. The medium sized city is well served by cinemas, theatre and restaurants. It is close to the National Exhibition Centre (for trade and leisure exhibitions and concerts) and within easy reach of Birmingham and London.

Contact
Prof A Jawaid
School of Engineering
Coventry University
Priory Street
Coventry
CV1 5FB
United Kingdom

Tel (+44) 1203 838071
                838012
Fax (+44) 1203 838082
Email enx916@cov.ac.uk

£2,490	
£7,300	
250 (PG)	
–	
60 : 40	
ESF/SRB	
–	
Throughout the year	
Extensive multimedia	
General/CAE facilities	
City centre	
£60–£95/£90–£132	
29 September 1997	
20 miles (Birmingham)	
1 hour 20 mins	
100 miles (160 km)	

Imperial College
The Centre for Composite Materials

Contact
Academic and other information
Mr F L Matthews

Tel   (+44) 171 594 5084
Fax   (+44) 171 594 5083
Email f.matthews@ic.ac.uk

Registration form from Admissions Office (Registry)
Imperial College
Exhibition Road
London SW7 2AZ

Tel   (+44) 171 594 8001
Fax   (+44) 171 594 8004

UNITED KINGDOM

Taught Courses

MSc IN COMPOSITE MATERIALS (ONE YEAR FULL-TIME; TWO OR THREE YEAR PART-TIME)
The full-time course is six months of lectures and coursework followed by six months of project work. Lectures are divided into about a dozen subject areas which include: Fibres; Matrices; Interfaces; Composite systems; Micro and macromechanics; Mechanical performance; Analytical techniques; Production methods; Mechanical testing and Design.

The laboratory programme involves fabrication, testing and analysis of a typical composite.

Other coursework includes a technical essay and a design project.

The research project is often carried out in collaboration with industry and can involve spending time at the sponsor's organisation. Recent project topics have included work on: Ceramic materials for artificial joints; Space craft materials; Repair of composite structures; Methods for bonding; Resin transfer moulding; Production of sandwich panels; Characterisation of metal matrix composites; Analysis of woven composites.

Research Programmes

MPHIL/PHD (TWO – THREE YEARS)
Staff in the Centre have a wide range of research interests. Research is currently being carried out in the following topic areas:
– Polymer Matrix Composites: abrasion and interfaces, polymer blends, adhesion, joining and repair, failure analysis and fracture mechanics, impact, fatigue, non-destructive testing, structural modelling, finite element analysis, test method development
– Metal & Ceramic Matrix Composites: production processes, microstructural analysis, mechanical properties

General Departmental Information
Imperial College is internationally recognised as the leading UK institution for all aspects of scientific, engineering and medical research. There is a large postgraduate population with a high proportion from overseas. About 30% of students are accommodated in Halls of Residence, mostly on campus. The Centre for Composite Materials has about 40 academic members of staff and 120 researchers. Strong contacts with UK and international companies ensure that the vast majority of the research work is carried in collaboration with industry.

Outstanding Achievements of the Academic Staff to Date
In addition to their teaching responsibilities the academic staff of the Centre for Composite Materials are also highly active researchers. The Centre was awarded the highest rating in the latest Research Assessment Exercise carried out by the Higher Education Funding Council for England, which amongst other criteria measures excellence in research, publications and ability to obtain funding.

Academic staff members are also heavily involved in both the organisation of and active participation in major conferences; at senior levels in learned society activities and other national, international and government panels and institutions.

Special Departmental Resources and Programmes
The Centre for Composite Materials has access to very well equipped research laboratories and excellent computing facilities.

Outstanding Local Facilities and Features
The Imperial College campus is located in South Kensington and is surrounded by major museums, artistic and other cultural facilities and park land, all of which are within two minutes walk. London is one of Europe's major capital cities with excellent travel connections to all areas of the world. An overwhelming range of cultural and social activities are available throughout the year.

🏛	£2,430
$¥	£8,800
👥	20
🏫	1 : 4
%	25%
📖	Yes
🔬	Yes
📅	31 July 1997
📖	Yes
💻	Very extensive
?	City
🛏	£23–£75
🗓	Early October 1997
✈	15 miles (Heathrow)
🏛	Located in capital
🚌	Located in capital

Napier University
Mechanical, Manufacturing and Software Engineering

Taught Courses

MSc/Postgraduate Diploma in Information Technology (Eight Streams)
- Computer Aided Engineering
- Control Systems Technology
- Engineering Design
- Manufacturing Systems Engineering
- Mechatronics
- Multimedia Technology
- Software Engineering
- Systems Integration

MSc/Postgraduate Diploma in Quality Management

MSc/Postgraduate Diploma in Materials Technology

MSc/Postgraduate Diploma in Environmental Technology
The courses are tailored in such a way that they will develop people with the professional qualities that the business community demands. The Postgraduate Diploma is the six month taught element of the course. This is followed by a six month project to Master of Science Degree.

General Departmental Information
The Department has invested in the most up-to-date high-technology equipment and facilities to provide an environment and atmosphere that advances the understanding and learning of our students. The teaching methods used are intended to foster student centred learning with the emphases on the integration of theoretical and applied work.

Outstanding Local Facilities and Features
Napier University is located in Scotland's capital city. Edinburgh offers a wealth of recreational and cultural pursuits. The Department of Mechanical Manufacturing and Software Engineering is located at the Merchiston Campus from which there is easy access to theatre, cinemas, museums, international conference centres, sports facilities and, of course Edinburgh's famous international festival.

Contact
Postgraduate Courses Administrator
Napier University
Deptarment of Mechanical, Manufacturing - Software Engineering
10 Colinlon Road
Edinburgh EH10 5DT
United Kingdom

Tel   (+44) 131 455 2301
Fax   (+44) 131 455 2264
WWW
http://www.mmse.napier.ac.uk

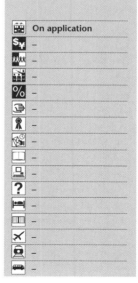

On application

UNITED KINGDOM

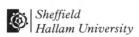

Sheffield Hallam University

Contact
Dr Mike Bramhall
Head of Postgraduate Studies
Sheffield Hallam University
City Campus
Pond Street
S1 1WB Sheffield
United Kingdom

Tel   (+44) 114 253 3377
Fax   (+44) 114 253 3433
Email M.Bramhall@Shu.ac.uk
WWW
http://www.shu.ac.uk/

Taught Courses
MBA Industrial Management; MSc/PGDip/PGCert Advanced Engineering; MSc Electrical and Electronic Engineering; MSc/PGDip/PGCert Engineering with Management; MSc/PGDip/PGCert Information Engineering

Research Programmes
The School of Engineering maintains a high profile in applied research across a range of topics in five generic areas. Research projects have been funded in Global Manufacturing, Tool Life Management, Novel Die Polishing, Corrosion Cracking in Metallic Alloys, and the electrical properties of thin films for sensor applications.
Over 40 research students are housed in purpose - built laboratories and work on research of over £2 million.
Main research themes include Mechanical Design, Electronics and Electrical Engineering, Manufacturing Engineering and Management, Structures Integrity and Control Systems.

General Departmental Information
Some of the most up to date laboratories and Engineering Equipment in European Higher Education are housed in the City Campus.

Outstanding Achievements of the Academic Staff to Date
Simulator of Tool Management Systems using DBMS – T.Perera; *Tool Cost Control*– T Perera R Tipple; *The role of knowledge based systems in the design of capital equipment* – G Cockerham

STAFF IN SCHOOL OF ENGINEERING
Product Introduction in PCB Design and Assembly – E K Lo; *Automatic in-process inspection during robotic PCBA rework* – E K Lo; *Experimental Study of the Printing of Solder Paste using the Metal Blade Squeegee System* – E K Lo; *Process modelling maps for solder paste printing* – E K Lo

PAPERS/BOOKS PUBLISHED BY SCHOOL STAFF
An investigation into the degradation and oxidation of certain hydrocarbon species in the exhaust manifold of an 81 engine – P Foss; *Applications of acoustics and vibrations in machinery health monitoring* – R B W Heng; *The acoustics of the passenger compartments of luxury vehicles* – R B W Heng; *Emission levels from Rover K16 Engine* – P W Foss; *A comparison of Hydrocarbon emissions from different piston designs in a 81 engine* – P W Foss; *Binary Cycle Turbine Powered Hybrid Vehicles* – P W Foss; *Domain Specific Minimum Environmental Impact Vehicles* – I Tranter, L Holloway, P W Foss; *Human Factor Influences on Effective CAD Implementation* – C Short; *Incorporative environmental principles into the design process* – D Clegg, I Tranter, G Cockerham; *Transputer control of a Flexible Manufacturing Coil* (FMC) – W Hales; *Robotic Kinematic Simulation for Sewer Renovation* – G Cockerham;
    £1.5 million Brite Euram Research Projects.
    About 15 Teaching Company Schemes are successfully operating.

Special Departmental Resources and Programmes
Wide range of State-of-the-art laboratories and computer hardware which houses some of the most advanced engineering software available.

Outstanding Local Facilities and Features
The Fifth largest city in Britain, Sheffield is centrally situated, with good motorway links. The Peak National Park is next to the city boundaries, with a range of superb scenery and leisure opportunities. Sheffield has world - class facilities for swimming, ice hockery, ski-ing and football. There is a wide range of theatres, galleries and the 12,000- seat arena for concerts.

🏛	–
$¥	–
👥	400 per year undergrad
🏢	–
%	20% : 80%
💬	Yes
🧍	–
📅	17 September 1997
🖥	Yes
🖥	Yes
?	City Centre
🛏	£30–£55 per week
📖	29 September 1997
✈	35 miles
🚗	2.5 hours
🚌	200 miles

University of Sussex
School of Engineering

UNIVERSITY OF

SUSSEX
AT BRIGHTON

Taught Courses
MSc in Digital Electronics and Parallel Processing Systems
Postgraduate Diploma in Digital Electronics and Parallel Processing Systems

Research Programmes
Biomedical Engineering; Electronic Engineering; Industrial Informatics and Manufacturing Systems; Physical Electronics and Instrumentation; Non-Linear Dynamics and Control; Power Electronic Control and Electrical Machines; Space Science; Structural Engineering; Thermo-Fluid Mechanics; VLSI and Computer Graphics

General Departmental Information
The Graduate Research Centre in Engineering (ENGG) is distinguished by the quality of both its research and its teaching. The multidisciplinary approach in teaching has been successfully carried across into research, with results that have consistently placed it high in the national research league tables.

ENGG achieved a Grade 4 (above average), in the General Engineering category, in both the 1988 and 1992 research assessment exercises conducted by the Higher Education Funding Council, with over 90 per cent of the faculty declared active in research. Its research success can also be measured by the consistently high level of research funding it has attracted over the years from industry and from the former Science and Engineering Research Council (SERC): the value of grants and contracts currently held exceeds £4 million.

Research Interests
Many faculty within ENGG have acquired an international reputation for their research, as evidenced by numerous invitations to participate in conferences throughout the world.

ENGG functions as a unified, general engineering unit. This non-departmental organisational structure has greatly facilitated the development of interdisciplinary research. The university's founders' belief that the most exciting developments eccur at the borders between traditional disciplines has certainly been verified in ENGG.

One example of this has been work in electromagnetic levitation, the original motivation for which lay in high-speed ground transportation without wheels. The continuing research needed to develop such systems embraces control theory and technology, power electronics, instrumentation and mechanical engineering. Latterly, this expertise has been applied more widely to magnetic bearings and to the active control of mechanical motor vibration.

Links with Industry
Research is motivated by social, economic and industrial needs. Strong links have been developed with local and national industry – not least because most practical problems requiring research also demand an interdisciplinary approach.

Relationships with industry are further exemplified in the university's developing technology park, where the companies involved are in engineering and have developed links with the university through contacts initially made via the School of Engineering.

Of particular concern to us are professional engineering standards. Engineers have obligations to the community which are regulated by the professional bodies. Our involvement is reflected in the senior positions held by our members of staff in engineering institutions such as the IEE.

Outstanding Local Facilities and Features
Brighton is a wonderful place to be a student – a brash, buzzing centre all the year round. Its eccentric and cosmopolitan atmosphere absorbs all kinds of people and leaves them with an enduring affection for the town, its community and its quirks. It is full of Sussex graduates who fell in love with the town and could not tear themselves away.

Brighton has one of the most active arts scenes in Britain, with an array of theatres, cinemas, concert venues and galleries. Brighton and Hove offer cuisine from around the world in more than 400 restaurants, and there are countless pubs, wine bars, cocktail bars and cafes. Sports facilities are excellent; two swimming pools, five golf courses, watersports, greyhound and horse racing, go-karting, ten-pin bowling, dry skiing and ice skating.

The highlight of the cultural year is the Brighton Festival in May. England's biggest art festival brings hundreds of international musicians, dancers and performers to Brighton for three weeks. The umbrella events – fringe theatre, jazz, classical music and comedy – add to the variety.

Contact
Dr L G Ripley
School of Engineering
University of Sussex
Falmer
Brighton
East Sussex
BN1 9QT
United Kingdom

Tel (+44) 1273 678358
Fax (+44) 1273 678452
Email l.g.ripley@sussex.ac.uk
WWW
http://www.sussex.ac.uk

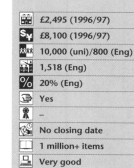

🏛	£2,495 (1996/97)
$¥	£8,100 (1996/97)
👥	10,000 (uni)/800 (Eng)
🏫	1,518 (Eng)
%	20% (Eng)
🖥	Yes
🎖	–
📅	No closing date
📚	1 million+ items
💻	Very good
?	Rural
🛏	£42+ (1996/97)
☐	October/January/April
✈	25 miles (Gatwick)
🚇	Less than 1 hour
🚌	65 miles

Brunel University
Department of Manufacturing and Engineering Systems

Contact
MSc courses
Mrs C Hirst
Postgraduate Admissions

Tel (+44) 189 527 4000
Ext. 2366
Fax (+44) 189 523 2806

Research
Professor B E Jones
Director BCMM
Department of Manufacturing
and Engineering Systems
Brunel University
Uxbridge
Middlesex UB8 3PH
United Kingdom

Tel (+44) 189 527 4000
Ext. 2514
Fax (+44) 189 581 2556
Email barry.jones@brunel.ac.uk

💷	£2,490 (October 1996)
💲	£8,195 (October 1996)
👥	230
👨‍🏫	1 : 3
%	20%
📖	Yes
🧑	Yes
📅	End of July 1997
📖	400,000 volumes
💻	Extensive network
?	Self-contained
🛏	£35–£55
📅	29 September 1997
✈	6 km
🕐	20 mins
🚌	24 km

Taught Courses

MSc in Advanced Manufacturing Systems
Modern manufacturing is changing faster than ever before. This course deals with the changing technologies, methodologies and strategies of manufacturing as well as providing a firm grounding in the basic principles. Subjects include: robotics and automation, computer integrated manufacturing, quality systems, production management, manufacturing measurement, manufacturing systems design, computer aided design and manufacturing strategy. *Note:* This course is available by distance learning. For entry a first degree in an appropriate subject. Other qualifications with experience may be considered.

MSc in Industrial Measurement Systems
An advanced multi-disciplinary programme in modern metrology and instrumentation covering basics of measurement science and technology, sensors, transducers and actuators, optical and mechanical metrology, automatic inspection and quality engineering, intelligent and distributed instrumentation. Excellent modern equipment resources are available in the Centre for Manufacturing Metrology. Modules are provided by experts from the Department of Manufacturing and Engineering Systems, Department of Physics and nearby National Physics Laboratory. A five-months research project forms part of the course. Entry requirements: a first degree in Engineering, Physical Sciences, Mathematics, or Computing.

Research Programmes

– **Measurements Systems – Manufacturing Metrology, Sensors and Quality Engineering:**
Including automatic inspection, condition monitoring, quality management, surface measurement, weighing systems.

– **Manufacturing Systems – Intelligent Design and Manufacture**
Including geometric modelling and design, intelligent automation, manufacturing strategy and management.

– **Engineering Systems**
Including bioengineering, flow measurement, solid mechanics and structures, electrostatics clean-up technology, simulation, modelling and optimisation, water systems monitoring and control software, internal combustion engines.
Most projects will be undertaken in collaboration with industrial companies or independent research organisations. EPSRC and University research studentships available. Applicants should have or expect to obtain I or II (i) honours degree in a physical science, engineering or technological subject.

General Departmental Information
This highly rated research department of general engineering integrates a range of multi-disciplinary engineering and management specialisations, includes the Brunel Centre for Manufacturing Metrology (BCMM), is a part of the University's Centre for Geometric Modelling and Design, and has established a Teaching Company Centre. It is strongly involved with an EngD programme in Environmental Technology. The department has strong research links with industry at home and abroad.
The Graduate School in the Department comprises 230 postgraduates (including 100 distance learning students) and there is an active programme of research training and seminars for research students. Research Council and departmental graduate studentships are available.

Outstanding Achievements of the Academic Staff to Date
The value of research grants and contracts held at any time averages at about £3 million with substantial funding being provided by the Research Councils and the European Union as well as industrial and other sources.
All researchers publish their work on a regular basis and over 250 papers have originated from the department in the last three years, many of which have appeared in international journals.

on the net http://www.editionxii.co.uk

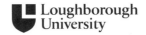

Loughborough University
Manufacturing Engineering

Taught Courses
The Department offers a modular full-time or part-time Master's course in Computer Integrated Manufacture. The course aims to produce engineers with the requisite technological knowledge and the management skills to integrate state of the art technological developments into the manufacturing system.

The University provides pre-sessional courses of up to four weeks for students who have not studied in United Kingdom previously. An English language support unit offers day and evening classes and a self-study language laboratory.

Research Programmes
Higher degrees by research (PhD and MPhil) may be studied full-time at the University or part-time within industry.

The major research areas are:
- CAD
- CAM
- CIM
- Electronics manufacture
- Manufacturing organisation
- Manufacturing processes
- Robotics, Machine control
- Systems' integration

General Departmental Information
The Department has large well equipped research laboratories for both conventional and electronics manufacturing. It is well equipped with over 60 Sun workstations which support its research activity.

Outstanding Achievements of the Academic Staff to Date
The Department is well known for the excellence of its research activities in collaboration with industry. 70% of its total income comes from this.

Outstanding Achievements of the University
The University received the Queen's Anniversary Prize for Higher and Further Education 1994 University.

Special University Resources and Programmes
The University has outstanding sports facilities – some of the most extensive in Britain which have been the training ground for many international athletes.

The Loughborough Students Union has over 13,000 members and is on of the most active in the country.

Outstanding Local Facilities and Features
Stamford Hall Theatre; Theatre Royal in Nottingham.

The Haymarket Theatre in Leicester; Loughborough Annual Fair; The Carillon in Queens Park; Golf Course in Shelthorpe; Bradgate Park; Beacon Hill.

"Advancing knolewdge through teaching, learning and research"

Loughborough University

Contact
G L Wiles
Loughborough University
Loughborough
Leicestershire LE11 3TU
United kingdom

Tel (+44) 150 922 2922
Fax (+44) 150 926 7725
Email G.L.Wiles@lboro.ac.uk
WWW
http://info.lboro.ac.uk/home.html

£2,490
£7,760
9,999 (total)
–
53%
6 available
As above
September 1997
Yes
Yes
Residential
£32–£62
30 September 1997
8 miles
2 hours
90 miles

**HERIOT-WATT
UNIVERSITY
*Edinburgh***

Heriot-Watt University
Civil and Offshore Engineering

Contact
Professor J Wolfram
Department of Civil and
Offshore Engineering
Heriot-Watt University
Edinburgh EH14 4AS
United Kingdom

Tel   (+44) 131 451 3142
Fax   (+44) 131 451 5078

Taught Courses
The Department offers a range of higher degrees relevant to the engineering profession and attractive to both students and prospective employers in the civil engineering, offshore and environmental industries. Studentships are available from the UK Engineering and Physical Sciences Research Council (EPSRC), Marine Technology Directorate Ltd, European Social Fund, and local authorities and industry. Home and overseas applications are welcomed.

MSc POSTGRADUATE DIPLOMA
– Geotechnical Engineering (with Glasgow University)
– Offshore Engineering
– Reliability Engineering and Safety Management
– Structural Engineering
– Subsea Engineering
– Subsea Engineering and Underwater Technology (with Cranfield Inst. of Tech.)
– Water Resources Engineering Management (with Glasgow University)

Research Programmes

RESEARCH LEADING TO THE DEGREES OF MPHIL AND PHD
– Estuarial and Coastal Engineering
– Fluid Loading and Instrumentation
– Geotechnical Engineering
– Marine Technology
– Materials Engineering
– Offshore and Pipeline Engineering
– Reliability Engineering
– Structural Engineering
– Systems Engineering
– Water Resources Management and Environmental Modelling

Special Departmental Information
The Department offers a rich combination of research and educational strengths, providing a range of soundly based programmes and an industrially relevant, strategic, and fundamental research capability for both the civil and offshore industries. The Department was assessed Grade 5 (the top rating) in the most recent UK universities' Research Assessment Exercise, one of only eight Civil Engineering departments in the United Kingdom, and the only department in Scotland, to do so.

General Information
Heriot-Watt University is well established. It was founded in 1966 as a technological university, but its origins can be traced back to 1821. The University has recently relocated to its new and expansive, but focused, parkland campus on the western edge of Edinburgh. A huge investment in new buildings and modern amenities has provided a stimulating high-quality educational environment with first class facilities for teaching, research and recreation.

Special Departmental Resources
The Department enjoys excellent laboratory and computing facilities, which range in scale from: a recently completed 12m by 12.4m wave basin with a depth of 3m (dropping to 5m); full-scale testing rigs for offshore pipelines; laser measurement of fluid flow; a.c. impedance spectroscopy facilities for investigating micro- and macro-properties of cementitious systems; corrosion monitoring facilities; to a range of powerful workstations for modelling large scale structural, geotechnical and hydrodynamic behaviour. Academic staff publish their research widely and represent the profession on various national and international professional bodies, editorial boards and conference committees.

Outstanding Local Facilities
Much of the character and architecture of the historical city of Edinburgh has been preserved, whilst a large technological industry has been allowed to thrive. There is a lively, cosmopolitan atmosphere to the city, with music, drama, and the arts well represented throughout the year and highlighted by the annual international festivals which attract people from all over the world.

	£2,430
	£5,993–£7,833
	–
	25 : 115
%	40:75
	Yes
	Yes
	September 1997
	Yes
	Yes
?	Modern
	£33–£40
	October 1997
	5 miles
	–
	608 km

on the net   http://www.editionxii.co.uk

University of Glasgow
Department of Mechanical Engineering

UNIVERSITY
of
GLASGOW

Taught Courses

MSc in Design Technologies
A modular taught part-time MSc in Design Technologies is under consideration which capitalises on design research and our expertise in CATIA.

MSc in Building Services Engineering
The part-time MSc in Building Services Engineering (GAITES) has been discontinued.

Research Programmes
Research topics include: Composite and ceramic materials; Hybrid optical computing; Laser materials processing and surface engineering; Fracture mechanics; Adhesive bonding; Holography in manufacturing; Membrane technology; Corrosion; Clean rooms; Control; Neural networks; Evolution of the design environments and communications; Gas dynamics and Aero engine performance.

General Departmental Information
The department offers a multi-disciplinary research environment for 40 to 70 researchers. Advanced engineering research topics feature strongly in one of the first engineering research centres in the world, and students from all over the world have benefited. The department enjoys considerable research contracts, including EC funded programmes.

Outstanding Achievements of the Academic Staff to Date
Internationally recognised authority in the quoted research areas.

Special Departmental Resources and Programmes
The department has outstanding computational infra-structure with the provision of network services for a heterogeneous distributed environment of UNIX workstations, enabling computationally intensive programmes to be developed in Gas Turbine Modelling and Elastic-Plastic Fracture Mechanics and large database handling in Design Information Systems. This infrastructure extends to facilities for data capture and processing for Control. Computer controlled experimental facilities including seven servo hydraulically controlled test machines with load capacities from 100 to 1,000 kN are central to work on adhesives and fracture mechanics. The Department is well equipped with optical microscopy, holographic surface analysis and reflectance spectroscopy facilities, and image processing systems. Experimental Laser research is founded on a strong equipment base including: a 1 kW CO_2; a 400W Nd: YAG; a 10W Ar ion; and a 20W laser diode array. A gantry system provides the capability for industrial scale laser processing, using a fibre-optic beam system which delivers 400W of Nd: YAG radiation via an IBM SCARA 7540 robot. In addition it supports scanning electron microscope and analysis facilities for materials evaluation in conjunction with Geology.

Outstanding Local Facilities and Features
- Glasgow justifiably earned the title "The European City of Culture"
- In 1998, Glasgow is The European City of Architecture and Design
- This friendly city has a lively cultural climate – and all the weather you will need!

Contact
Professor Brian F Scott
Mechanical Engineering
James Watt Building
University of Glasgow
Glasgow
G12 8QQ
United Kingdom

Tel (+44) 141 330 4317
Fax (+44) 141 330 4343
Email r.murray@Mech.gla.ac.uk.

🏫	£2,490
💲	£8,370
👥	16,000
🏢	–
%	55
📖	–
👤	–
🗓	–
🖥	Excellent
🖥	Excellent
?	City
🛏	£45
	–
✈	13 miles
🚂	50 mins to Edinburgh
🚌	65 km to Edinburgh

THE
UNIVERSITY
OF HULL

University of Hull
Engineering Design and Manufacture

Contact
Dr R D James
Department of Engineering
Design and Manufacture
School of Engineering
and Computing
University of Hull
Hull
HU6 7RX
United Kingdom

Tel   (+44) 148 246 6222
Fax   (+44) 148 246 6533

The Department of Engineering Design and Manufacture was established in 1979 in response to the Finniston report on engineering education. Since that time, its research reputation has developed rapidly and it was one of very few engineering departments to obtain the highest rating in the 1992 research selectivity exercise.

Links with industry, in the UK and abroad, are well-established and most of the senior academic posts in the Department are supported by industrial companies. The Department has links with the international academic community and, in particular, has formal exchange programmes with Yanshan University in the P R of China and with Yonsei University in South Korea.

Taught Courses
The Department offers a one-year full-time taught MSc course in Advanced Manufacturing and Materials, which builds on its research excellence in these key areas. The course involves a combination of formal lectures and project work and assessment is based on written examinations and a dissertation.

A second class honours degree, or better, is the normal entry requirement.

Research Programmes
The Department accepts suitably qualified students on full-time and part-time PhD, MPhil and MSc research degree programmes in the following areas:
– Design for Economic Manufacture
– Intelligent Knowledge-based Systems
– Surface Engineering
– Wear of Ultra-hard Materials
– Mechanical Properties of Engineering Ceramics
– Finite Element Analysis
– Low Dimensional Structures
– Tribology
– Laser-based Inspection
– Acoustic Modelling

Each research student works as part of one of the established groups within the Department and has access to well-equipped laboratories. Essential support for project work is provided by technicians with facilities for the design and construction of specialist experimental equipment.

There is a formal programme of research seminars within the Department and it is a requirement that all postgraduate students attend these regularly. Each PhD student is required to give a seminar on the research work they have undertaken before being allowed to submit a thesis for examination.

General information
Founded in 1927, the University is situated on an attractive campus approximately 4km from the centre of the historic city of Kingston upon Hull and is within easy reach of unspoiled countryside and coast.

£2,430
£7,950
1,250 (Postgraduate)
1 : 3
35 : 65
Yes
–
July (Crse)/Rsch any tme
Yes
Yes
City fringe
£30–£35
Sept (MSc)/Rsch Flex
20 miles
2.5–3 hours
220 miles

Promoting Excellence in Education and Research

on the net **http://www.editionxii.co.uk**

Loughborough University of Technology
Department of Mechanical Engineering

Taught Courses

MSc IN DESIGN OF MECHATRONIC PRODUCTS

This programme provides an integrated course in both the skills of engineering design and the necessary technical background for a true mechatronic engineer (i.e. the integration of mechanical elements, microprocessors, software, and control) is growing around the world. Hence engineers/technologists skilled in this area will be in demand in industry as we approach the new century.

MSc IN ENGINEERING DESIGN

This programme is the largest of the three courses and is popular with postgraduate students from all around the world. It is based upon design methodologies skills for the professional engineer. This is expanded upon in areas such as marketing, CAE, industrial design and human factors engineering. Successful postgraduate designers are in much demand all around the world.

MSc IN MECHATRONICS AND OPTICAL ENGINEERING

This programme is a hybrid incorporating most of the technical skills of the Mechatronics Products course, but replacing design methodologies modules with modules in optics, vision systems and laser engineering. This is primarily a technical based course, generating a practitioner ready to provide solutions for the mechatronic and optical needs of industry.

Course Information

All programmes are modular in nature and are available on a full-time or part-time study basis.

Apply

Application forms and further information about taught MSc courses can be obtained from the Postgraduate Courses Admissions Tutor and /or Administrator.

Research Programmes

MPHIL AND PHD BY FULL-TIME OR PART-TIME RESEARCH IN THE FOLLOWING AREAS

Engineering Design Methods and Management; Medical and Orthopaedic Engineering; Mechatronics; Dynamics and Control; Strength of Materials; Thermodynamics and Fluids; Internal Combustion Engines; Textiles Machinery; Optical Engineering; High Speed Machinery.

Apply

Application forms and information about research opportunities can be obtained from the Research Co-ordinator. Research enquiries are welcome at any time.

General Information

The Engineering School at Loughborough University of Technology is one of the largest in the United Kingdom. The town of Loughborough offers a friendly community of 50 000 people in a beautiful part of central England with good road and rail networks to other regions. The University is home to 12 000 students, many of whom are from overseas.

Outstanding Local Facilities and Features

Rural East Midlands, close to Charnwood Forest, and the Peak District. Local theatre and culture in Nottingham and Leicester. Close to the M1 motorway giving good access to the United Kingdom generally.

Contact

Helen Sankey
(Administration)

Paul King
(Admissions Tutor)

Tel (+44) 150 922 3176
Fax (+44) 150 922 3936
Email p.d.king@lut.ac.uk.

Menim Acar
(Research Coordinator)
Tel (+44) 150 922 3218
Fax (+44) 150 922 2934
Email acar@lut.ac.uk.

Engineering Design Institute
Loughborough University
of Technology
Department of Mechanical
Engineering
Ashley Road
Loughborough
Leicestershire
LE11 3TU
United Kingdom

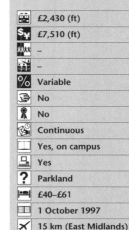

🏫	£2,430 (ft)
$¥	£7,510 (ft)
👥	–
🏛	–
%	Variable
🖥	No
👤	No
🕐	Continuous
🏠	Yes, on campus
💻	Yes
?	Parkland
🛏	£40–£61
📅	1 October 1997
✈	15 km (East Midlands)
🚉	1.20 hour
🚌	160 km

UMIST

University of Manchester Institute of Science and Technology
Mechanical Engineering Department

Contact
Dr M Robinson
Mechanical Engineering
Department
UMIST
Sackville Street
PO Box 88
Manchester M60 1QD
United Kingdom

Tel   (+44) 161 200 3754
Fax   (+44) 161 200 3755
Email mepgadm@umist.ac.uk
WWW
http://www.me.umist.ac.uk

💰	£2,490
💲¥	£8,100
👥	145
📊	1 : 3.5
%	80
💱	In certain cases
🧍	–
📅	August 1997
📖	Good
💻	Excellent
?	City centre
🛏	£60
📅	22 September 1997
✈	10 km
🚗	2.5 hours
🚌	300km

Taught Courses
The Department offers four MSc courses by Examination and Dissertation, each lasting 12 months full-time. Fuller information is available in the Departmental Postgraduate Prospectus.

COMPUTATIONAL AND EXPERIMENTAL STRESS ANALYSIS
This course includes the following topics – Elasticity; Plasticity; Impact; Composites;, Finite Elements; Numerical Methods; Experimentation and Instrumentation; Shell Structures; Metal Forming.

MECHANICAL ENGINEERING DESIGN
Similar to above but with Design topics replacing Experimental Techniques.

THERMAL POWER AND FLUIDS ENGINEERING
This course covers the following subjects – Experimental Methods; Fluid Mechanics; Heat Transfer; Thermodynamics; Numerical Analysis; Computational Fluid Dynamics; Power Generation. A particular feature is the emphasis given to Experimental Techniques.
    There is also a new MRes in Computational and Experimental Fluids Engineering, providing advanced training in the title topics and a preparation for PhD level research.

ADVANCED MANUFACTURING TECHNOLOGY AND SYSTEMS MANAGEMENT
The course contains core modules including Manufacturing Systems; Processes. Management and Automation; CAD and CAM, Metrology, Machine Tool Performance, with further optional modules in Manufacturing Systems, Management Processes, and Computer Aided Modelling of Materials.

Research Programmes
The Department has an extensive research activity based on the four Divisions of Applied Mechanics; Thermodynamics and Fluid Mechanics; Polymer Engineering; Manufacturing. The main areas of interest include Computational Solid Mechanics; Composites; Impact Loading; Metal Working; Structural Plasticity; Computational Fluid Dynamics; Heat Transfer; Internal Combustion Engines; Polymer Processes and Manufacturing Systems; Computer Aided Design and Manufacture; Machining Processes; Materials in Design and Manufacture.

General Departmental Information
The Department has a long history of postgraduate education, research, and active collaboration with industry and Government Departments. It has been running MSc taught courses for over thirty years. There are 39 academic members of staff and about 145 full-time postgraduate students. Half of these are registered for the degree of PhD, the remainder for MPhil, MRes, MSc or the Postgraduate Diploma.

Outstanding Achievements of the Academic Staff to Date
The Department has an outstanding record of research achievement, and related expertise feeds directly into postgraduate courses. It is one of only five Mechanical Engineering Departments in the UK awarded "5A" research rating – the highest possible – in the 1992 Research Selectivity Exercise, and it is confident of retaining this grade in the 1996 Exercise.
    Academic staff publish approximately 140 papers per year, mostly in international journals and conferences. Members of the Department hold ten higher doctorates and 24 Fellowships, among them one Fellowship of the Royal Society and three Fellowships of Engineering, have contributed to the organisation of some 58 international conferences over the past four years and are members of 21 editorial boards of journals.

Special Departmental Resources and Programmes
The Department has outstanding laboratory facilities served by its own fully equipped workshops and technical staff. Computing capabilities are also excellent with many numerical analysis software packages available. Present areas of particular interest are Impact Mechanics and CFD applied to Turbulent Flows.

Outstanding Local Facilities and Features
Manchester has excellent cultural and sporting facilities, for example the Royal Exchange Theatre, The University Museum, The Hallé and BBC Philharmonic Orchestras, and a lively club scene. It is well-known for its two local soccer teams and will host the Commonwealth Games in 2002. Nearby beauty spots include the Lake District, Peak District, the Roman city of Chester, and the castles and mountains of North Wales.

Aston University
Mechanical and Electrical Engineering

Taught Courses

MSc/PGDip in Risk Management and Safety Technology

This modular programme is designed to meet the growing demand for health and safety education which has arisen from an increased awareness of the importance of risk assessment.

Emphasis is placed upon the technical engineering issues associated with risk management.

Students can also study discrete modules in the programme. These can be accumulated to achieve a Postgraduate Diploma and further research and submission of a dissertation may lead to an MSc.

Research Programmes

There are two major research groups in the Department.

Dynamics, Control and Vibration Group have several major EPSRC/DTI projects funded under the Design of High Speed Machinery LINK initiative, concerned with maximising dynamic performance of machines incorporating servo motors. Work on the efficient simulation of servo-machine systems is being advanced, along with the development of motors and drives of ever higher dynamic capabilities. In structural dynamics, the group is contributing to developing methodologies for updating model parameters and assessing model quality, reducing models and computing efficiently resonances of finite periodic and composite structures. The group is applying modelling and identification techniques to problems in rotor-bearing-foundation systems, particularly the detection of faults.

The Risk Assessment and Management Group is the leading centre for advanced study of occupational health and safety in the UK and collaborates with the chemical, nuclear, energy, transportation industries and with HSE. The group focuses on socio-technical approaches to risk control in engineered systems. Research covers communications and decision-making (within the notion of 'safety culture') and development of analytical frameworks for examining organisations. From an engineering perspective, research includes work on intrinsically safe designs and mitigation of confined explosions, development of risk assessment methodologies, and quantitative methods for risk estimation for industries now relying on qualitative methods.

Research is also undertaken in electrical power utilisation, cutting and drilling technology and in the processing and properties of engineering materials.

General Departmental Information

The Department is equipped with excellent research facilities, and benefits from Aston's leading national position in IT resources. Although the academic interests of the Department's research groups differ, the underlying objective of each group is the same: to undertake work of the highest quality in areas of national and international significance.

The members of our research groups work closely with industry, including companies based at Aston Science Park.

Outstanding Achievements of the Academic Staff to Date

– Dr J E T penny is co-ordinator of a Brite-EuRam project involving seven partners in five countries
– Prof M R Hayns, an internationally known expert in Nuclear Safety was awarded an OBE for his services to nuclear safety
– Dr S Garvey has been awarded 3 EPSRC Research Grant since 1994 in machine dynamics

Social Department Resources

Machine Control and Drive Research Laboratory costing £3/4 million provides a unique location where lectures and seminars to students and industrialists alike can be carried out, supplementing theoretical analysis with practical experiments.

Rotor Dynamics Research Laboratory contains a comprehensively instrumented rotor test rig in which the rotor is supported by four oil film bearings on flexible foundations. A high speed test rig, running at 100,000 rev/min in magnetic bearings, is being commissioned.

Outstanding Local Facilities and Features

Birmingham is an international city with first class theatres; Clubs; a Museum and Art Gallery and has recently acted as one of the host cities for the Euro Cup 96 football competition. Festivals held include the International Jazz Festival; a Film and TV Festival and a Comedy Festival. Shopping facilities include sophisticated precincts, bookshops and open and under cover markets.

Contact
Dr J E T Penny
Aston University
Aston Triangle
Birmingham B4 7ET
United Kingdom

Tel   (+44) 121 359 3611
       Ext 4280
Fax   (+44) 121 333 5809
Email MECHELEC@aston.ac.uk

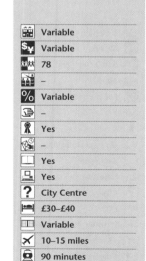

Variable
Variable
78
–
Variable
–
Yes
–
Yes
Yes
City Centre
£30–£40
Variable
10–15 miles
90 minutes
130 miles

Imperial College
Earth Resources Engineering

Contact
Dr Richard Dawe
Centre for Petroleum Studies
Earth Resources Engineering
Department
Imperial College
London SW7 2AZ
United Kingdom

Tel   (+44) 171 594 7325
Fax   (+44) 171 594 7444
Email r.dawe@ic.ak.uk

Taught Courses

PETROLEUM ENGINEERING

The MSc course in Petroleum Engineering provides a 12-month 'conversion' from other engineering and science based degree foundations into the specialities of Petroleum Engineering. The course should enhance the employment prospects in the oil industry.

Petroleum engineering involves the application of earth and physical sciences to the evaluation and exploitation of natural hydrocarbon resources. The dominant problems of the petroleum engineer are those of flow and equilibrium in-porous media, in vertical well bores, in surface pipelines and in primary process equipment. The complexity of the hydrocarbon fluids, and the geological strata involved in flow in reservoirs and production systems raises problems requiring sophisticated numerical techniques for their solution. In the practical field, drilling and production engineering continually pose new engineering problems requiring engineered solutions.

The competition and demands are such that entrants to the Imperial College MSc course must have at least a good second class Honours degree in engineering or science. Lesser formal qualifications are acceptable only when applicants have good relevant experience and first class references. In all cases the conditions for University of London registration must be met.

University level mathematics and a working knowledge of programming in a scientific language are also highly desirable.

A number of Engineering and Physical Sciences Research Council Advanced Course Studentships, industrial scholarships and bursaries are available to candidates for this course and are awarded on the results of competitive interviews held in the late spring.

Research Programmes

The Department's research is organised into co-ordination areas of individual research groups, concerned principally with the science and engineering of characterisation and extraction of primary mineral and energy resources and of the related environmental problems and solutions. Research students and research assistants register for the higher degrees MPhil and PhD of the University of London plus the Diploma of the Imperial College (DIC). They are expected to participate in all research training activities of their research group and to develop their communication skills through preparation of papers and participation in seminars and colloquia. Prospective researchers are invited to request the Department's Research Review which describes current and recently completed research projects.

Outstanding Local Facilities and Features

All the usual (and unusual) London facilities and features are on your doorstep!

💰	£2,540
$¥	£9,100
👥	6,000 (ug + pg)
	30–40%
%	For home students
	Yes
	–
	Up to starting date
	Excellent
	Excellent
?	City
	£80
✈	October 1997
✈	13 miles
	You are in it
	You are in it

University of Glasgow
Naval Architecture and Ocean Engineering

UNIVERSITY
of
GLASGOW

Taught Courses

DIPLOMA IN NAVAL ARCHITECTURE AND OCEAN ENGINEERING
A one year course for graduates in Engineering or Science that provides a background for a career or research in these areas.

MSc IN OFFSHORE ENGINEERING
Taught in conjunction with Heriot-Watt University this one year course would suit graduates interested in a Naval Architectural and/or Structural role in the Oil industry.

Research Programmes

MSc OR PhD RESEARCH
The main research areas are:
–   Computational and experimental hydrodynamics, e.g. Dynamics of moored deep water offshore platforms; Hydrodynamics of manoeuvring ships; Conceptual design of advanced marine vehicles; Deep water catenary riser systems; Heavy-weather oil slick containment systems
–   Structural strength and reliability based model development e.g. Statistical modelling of metocean parameters and environmental forces; Reliability based strength formulations for ring and stringer stiffened cylinders; Reliability based inspection planning. Novel design of fast ships, ferries, tankers and bulk carriers. Stability of ships: Particularly Ro-Ro ferries
–   Composite structures and adhesives: e.g. application to offshore and fast marine vehicles

General Department Information
The Department's academic staff currently consists of the Professor, two Senior Lecturers and three Lecturers. In addition there are seven technical staff at the Hydrodynamics Laboratory, including a computer manager. A departmental library is available for the day-to-day needs of postgraduate students. The Department has Dec and Sun workstations, as well as many PCs and Macintosh computers. All computers are networked with each other and with the central Faculty of Engineering workstations.

Outstanding Achievements of the Academic Staff to Date
The following members of the Department's staff represent the United Kingdom on international committees:
–   Professor Nigel Barltrop, International Standards Organisation (ISO)
–   Dr. Purnendu K. Das, International Ship and Offshore Structures Congress (ISSC)

BOOKS WRITTEN OR EDITED BY STAFF MEMBERS
Nigel Barltrop, *Fluid Loading on Fixed Offshore Structures*, HMSO, 1990; *Dynamics of Fixed Marine Structures*, MTD/Butterworth-Heinemann, 1991; and *Floating Structures: A Guide to Design and Analysis*, MTD/OPL, 1996. Purnendu Das, editor of the Integrity of Offshore Structures series, published in 1987, 1990, and 1993.

Special Departmental Resources and Programmes
The Department, in its Hydrodynamics Laboratory, has the largest towing/wave tank of any British University. The laboratory is equipped with advanced measuring equipment and data logging and analysis systems, and houses its own model making, electronics and mechanical engineering workshops. Resistance, manoeuvring, seakeeping, collision and slamming impact studies are carried out at the Hydrodynamics Laboratory. Strength, fatigue and material tests are carried out in the Civil and Mechanical Engineering Laboratories.

Outstanding Local Facilities and Features
Glasgow is the home of Scottish Opera, Scottish Ballet, the Scottish National Orchestra and the Scottish Exhibition and Conference Centre. The City's public libraries are the largest in Europe. Among the many art galleries and museums is the award-winning Burrell Collection, one of Scotland's leading tourist attractions. Ideal for the outdoor pursuits enthusiast, the Lochs and Mountains of the Scottish Highlands are just a 40 minutes drive from Glasgow.

Contact
Professor Nigel Barltrop
Department of Naval
Architecture and Ocean
Engineering
James Watt Building
University of Glasgow
Glasgow G12 8QQ
United Kingdom

Tel   (+44) 141 330 4322
Fax   (+44) 141 330 5917
Email nbarltrop@eng.gla.ac.uk

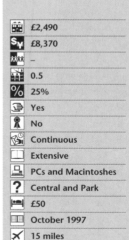

£2,490	
£8,370	
–	
0.5	
25%	
Yes	
No	
Continuous	
Extensive	
PCs and Macintoshes	
Central and Park	
£50	
October 1997	
15 miles	
1 hour (Edinburgh)	
50 miles (Edinburgh)	

University of Strathclyde
Ship and Marine Technology

Contact
Professor Chengi Kuo
Department of Ship
and Marine Technology
University of Strathclyde
100 Montrose Street
G4 0LZ Glasgow
United Kingdom

Tel   (+44) 141 552 4400
Fax   (+44) 141 552 2879
Email c.kuo@strath.ac.uk

Research Programmes

MPHIL BY RESEARCH IN ADVANCED MARINE TECHNOLOGY
Research Topics:
–   Cable Dynamics: Offshore lifting; control and compensation systems
–   Fluid Loading: Computational fluid dynamics; behaviour of offshore structures
–   High Speed Craft: Dynamic loads; stability; safety
–   Hydrodynamics: Riser, pipeline and marine vehicle motions
–   IT Applications: Marine usage of telepresence and virtual reality
–   Materials: Polymer behaviour in the oceans
–   Safety: Safety case approach; performance measurement in industry
–   Ship Stability: Ro-Ro passenger ships; damaged stability; capsizing
–   Subsea Intervention: AUV/ROV design; hydrodynamic coefficients

General Departmental Information
The Department of Ship and Marine Technology has a proud history and an international reputation in naval architecture, ship production technology, offshore engineering and other areas of marine technology, including the design, performance and construction of fishing vessels, yachts and other small craft. Teaching began in 1882 and benefited considerably from close links with the local shipbuilding community. Almost a hundred years later, in 1976, the Marine Technology Centre was created through an initiative of the Science and Engineering Research Council (SERC), with specific attention directed towards technologies associated with offshore exploration and production of oil and gas. Numerous interdisciplinary research projects have resulted, with the major input coming from the Department of Ship and Marine Technology. Students of more than fifty nationalities have graduated from the Department.

Outstanding Achievements of the Academic Staff to Date
Some examples include:
–   Published textbooks in several areas include: computer application, business and safety, e.g. *"Business Fundamentals for Engineers"* (McGraw Hill)
–   Initiators of novel teaching approach by group-work involving team activities, performance assessment, presentation
–   Successful application of the 3-C educational philosophy that enables graduates to achieve a balance of competence, confidence and communication skills
–   International reputation as leading researchers in ship stability, maritime safety management; ship support for sub-sea work in offshore oil production
–   Good working relationship with over 20 institutions world-wide

Special Departmental Resources and Programmes
The Department has its own lecture theatres, extensive computing and study areas, and a large design office. The laboratory has a wide range of facilities including a seakeeping basin of dimensions 17m x 9m x 3m for omnidirectional wave and current studies, and a towing tank measuring 25m x 1.5m x 0.85m. Under the terms of a recent agreement, the Department has access, for teaching and research, to the Denny Ship Model Experiment Tank (92m x 7m x 2.7m). The University can also offer a large selection of other facilities, including materials testing and fluid mechanics laboratories, a wind tunnel, an advanced machine-tool laboratory and extensive IT facilities.

Outstanding Local Facilities and Features
Glasgow is the largest city in Scotland and one of the liveliest in Britain. It has undergone substantial 'urban renewal' in recent years, making it a most attractive place in which to live and study. Glasgow was chosen as the European City of Culture in 1990 and now offers an exceptionally wide range of social, cultural and sporting activities. Glasgow is bursting with fine architecture and renowned museums and yet still has space for a multitude of parks to justify its description as the 'dear green place'. The cost of living is still substantially lower than in many areas, particularly those in the southern UK. It is surprisingly easy to leave the city and explore the lovely Scottish countryside. Some of the best outdoor sports in Britain, such as sailing, boating, windsurfing, hill-walking, climbing and skiing, is within easy reach.

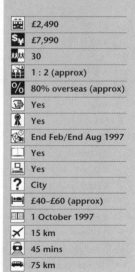

💷	£2,490
💲	£7,990
👥	30
🏫	1 : 2 (approx)
%	80% overseas (approx)
📖	Yes
👤	Yes
📅	End Feb/End Aug 1997
📋	Yes
💻	Yes
?	City
🛏	£40–£60 (approx)
🗓	1 October 1997
✈	15 km
🚉	45 mins
🚌	75 km

on the net   http://www.editionxii.co.uk

Imperial College
Earth Resources Engineering

Taught Courses

PETROLEUM ENGINEERING

The MSc course in Petroleum Engineering provides a 12-month 'conversion' from other engineering and science based degree foundations into the specialities of Petroleum Engineering. The course should enhance the employment prospects in the oil industry.

Petroleum engineering involves the application of earth and physical sciences to the evaluation and exploitation of natural hydrocarbon resources. The dominant problems of the petroleum engineer are those of flow and equilibrium in-porous media, in vertical well bores, in surface pipelines and in primary process equipment. The complexity of the hydrocarbon fluids, and the geological strata involved in flow in reservoirs and production systems raises problems requiring sophisticated numerical techniques for their solution. In the practical field, drilling and production engineering continually pose new engineering problems requiring engineered solutions.

The competition and demands are such that entrants to the Imperial College MSc course must have at least a good second class Honours degree in engineering or science. Lesser formal qualifications are acceptable only when applicants have good relevant experience and first class references. In all cases the conditions for University of London registration must be met.

University level mathematics and a working knowledge of programming in a scientific language are also highly desirable.

A number of Engineering and Physical Sciences Research Council Advanced Course Studentships, industrial scholarships and bursaries are available to candidates for this course and are awarded on the results of competitive interviews held in the late spring.

Research Programmes

The Department's research is organised into co-ordination areas of individual research groups, concerned principally with the science and engineering of characterisation and extraction of primary mineral and energy resources and of the related environmental problems and solutions. Research students and research assistants register for the higher degrees MPhil and PhD of the University of London plus the Diploma of the Imperial College (DIC). They are expected to participate in all research training activities of their research group and to develop their communication skills through preparation of papers and participation in seminars and colloquia. Prospective researchers are invited to request the Department's Research Review which describes current and recently completed research projects.

Outstanding Local Facilities and Features

All the usual (and unusual) London facilities and features are on your doorstep!

Contact
Dr Richard Dawe
Centre for Petroleum Studies
Earth Resources Engineering
Department
Imperial College
London SW7 2AZ
United Kingdom

Tel   (+44) 171 594 7325
Fax   (+44) 171 594 7444
Email r.dawe@ic.ak.uk

£2,540	
£9,100	
6,000 (ug + pg)	
30–40%	
For home students	
Yes	
–	
Up to starting date	
Excellent	
Excellent	
City	
£80	
October 1997	
13 miles	
You are in it	
You are in it	

University of Aberdeen
Safety Engineering Unit
Department of Engineering

Contact

Safety Engineering Unit
Department of Engineering
Fraser Noble Building
University of Aberdeen
King's College
Aberdeen AB24 3UE
United Kingdom

Tel (+44) 122 427 3895
Fax (+44) 122 427 2497
Email mjb@eng.abdn.ac.uk
WWW
http://www.eng.addn.ac.uk/

Taught Courses

The Safety Engineering Unit provides a one year programme in Safety Engineering and Risk Management leading to the degree of Master of Science (MSc). It may also be taken part-time over two or three years.

The programme is designed to provide advanced training for engineers wishing to specialise in Safety and Reliability Engineering. It includes courses in: Basic safety, reliability, risk concepts and safety legislation; Statistics and probability for safety, reliability and quality; Advanced methods for risk and reliability assessment; Quality control, risk management and human reliability; Process safety; Fire and explosion engineering; Software reliability and safety-critical systems; Reliability and safety of electrical/electronic systems; Reliability-based fatigue and fracture assessment; Reliability of structural systems. Students also study one of the following option courses: Offshore Engineering (including petroleum engineering, sub-sea technology and drilling technology); Ergonomics and Occupational Safety; Environmental Engineering. An industry-based project is undertaken during the last three months.

Applicants should have a good honours degree in engineering, or a science-based degree together with relevant industry experience.

A number of government and industrially funded bursaries are available covering fees and maintenance costs.

Research Programmes

The Unit is involved in a wide range of research. The main themes are: development of methods for the reliability assessment of engineering systems: response of structures to blast and impact loading; assessment of existing structures incorporating reliability updating; combustion and the control of combustion products; methods for risk assessment and safety management; human reliability; safety critical systems and software reliability; driver misperception in road accidents; computer-based learning; evaluation of safety legislation.

General Departmental Information

The Safety Engineering Unit is part of the Department of Engineering which has about 40 full-time academic staff and a complement of 35 technicians and other staff. At postgraduate level, in addition to MSc Programmes in Safety Engineering and Risk Management and Project Management, the Department has about 70 research assistants and students undertaking research in a wide range of engineering disciplines. Together with the subjects listed above, the main research areas include: Offshore Engineering; Laser Technology and Optical Engineering; Non-Linear Dynamics; Advanced Materials; Geotechnical Dynamics; Structures; Hydraulics and Sediment Transport; Isolated Electrical Power Systems and Power Electronics; Condition Monitoring; Intelligent Control; VLSI Design; Communications; Thin Films; and Nano-Technology. The Department is an associate department of the University's Oil and Gas Institute.

The Safety Engineering Unit was formed in 1991 and is supported by the Elf Enterprise Consortium. It has a core staff of five, together with research assistants and students. It is assisted in its teaching role by other academic staff from the Department of Engineering and other University Departments.

Outstanding Achievements of the Academic Staff to Date

P Thoft-Christensen and M J Baker, (1982), *Structural Reliability Theory and its Applications*, Springer-Verlag;
P Vas, (1992), *Electrical Machines and Drives, A Space Vector Theory*, Oxford University Press;
J C Jones, (1993), *Combustion Science: Principles and Practice*, Millennium Books;
J C Jones, (1996), *Topics in Environmental and Safety Aspects of Combustion Technology*, Whittles Publishing.

Special Departmental Resources and Programmes

The Department has excellent computer and laboratory facilities, and close connections with, and support from many of the companies based in Aberdeen and involved in the North Sea oil and gas industry.

Outstanding Local Facilities and Features

Aberdeen has a population approaching 250 000 and distinctive granite architecture leading to the name – the Silver City. It supports a thriving cultural life with regular symphony and chamber concerts, a fine Edwardian theatre and an exceptional museum and art gallery. The traditional industries have been joined over the last two decades by oil-related industries leading to its new reputation as the oil capital of Europe. Diverse sporting opportunities are available, ranging from golf to skiing, and climate is typically dry with average rainfall and sunshine similar to that of London, although some 2–3ºC cooler.

🏛	£6,150
💲¥	£7,920
👥	–
🎓	1 : 3
%	20%
📖	Yes
🧍	Yes
🕐	Continuous
📖	Main University library
💻	UNIX and PC networks
?	Historic site in Aberdeen
🛏	–
📅	Mid–September
✈	6 miles (10 km)
🚊	2.5 hours (Edinburgh)
🚌	125 miles (200 km)

on the net http://www.editionxii.co.uk

United States

University of Florida
Agricultural and Biological Engineering

Taught Courses
We offer advanced study leading to degrees in either the College of Engineering or the College of Agriculture. The degrees of Master of Engineering and PhD in the College of Engineering are intended for students who have completed an undergraduate degree in engineering while the Master of Science (College of Engineering) is intended for students who do not have an engineering degree but desire to advance their science backgrounds within the framework of an engineering discipline. The degrees of Master of Science and PhD through the College of Agriculture provide for scientific training and research in technical agricultural management.

Research Programs
Our research program covers four general areas:
- **Food and Bioprocess Engineering**: Postharvest engineering; Process microbiology; Heat and mass transfer in biological systems; Thermal processing; and Recycling systems for waste and protein production
- **Water Resources Engineering**: Irrigation; Drainage; Non-point pollution control; Watershed hydrology; Solid waste disposal; Groundwater hydrology; Waste management; Water reuse and Remote sensing
- **Production Engineering**: Machine systems and design; Aquaculture production systems; Safety; Structures and their environment; Pesticide application; Systems management; Automation and aquatic biomass production
- **Information Systems**: Electronic communication technology, especially CD-ROM, with special topical emphases in safety and energy, mathematical modelling of plant and animal systems; and Knowledge-based decision support systems

General Department Information
Our department is one of the largest Agricultural and Biological Engineering departments in the US with 27 professors on the main campus and seven other professors at the various education and research centers throughout Florida. The enrolment is 220 undergraduate students (in Agricultural and Biological Engineering and Agricultural Operations Management) and 40 postgraduates. We are part of both the College of Engineering (ranked in the top 20 best graduate programs in the US) and an outstanding College of Agriculture. The University of Florida is a member of the prestigious Association of American Universities, a very selective organization of 60 preeminent graduate research universities in North America.

Outstanding Achievements of the Academic staff to Date
Our current faculty has received seven College Teachers or Advisors of the Year awards.

Special Departmental Resources and Programs
Well equipped laboratories and computer facilities including multi-media. We also have access to research facilities in Food Science; Horticultural Sciences and other departments on campus.

The Colleges of Engineering and Agriculture and the University of Florida have excellent computer facilities.

The eight libraries at the University of Florida, with over three million catalogued volumes, form the largest information resource system in the state of Florida.

Outstanding Local Facilities and Features
Gainesville, with a population of almost 94,000, offers the best of small town friendliness and big city sophistication. It was rated the number one city (best place to live) in America by Money Magazine in 1995. UF students can attend performances at the Hippodrome State Theater and the Center of Performing Arts; visit the Florida Museum of Natural History, the Harn Museum of Art and exhibits at the University of Florida Gallery. The 2,000 - acre centrally located campus is 90 minutes from the Gulf of Mexico and the Atlantic Coast. North Florida climate offers the best of all seasons.

Contact
Graduate Co-ordinator
Agricultural and Biological Engineering
PO Box 110570
University of Florida
Gainesville
Florida 32611
USA

Tel (+1) 352 392 1864
Fax (+1) 352 392 4092
Email kvc@agen.ufl.edu
WWW AGEN.UFL.EDU

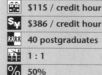 $115 / credit hour

 $386 / credit hour

40 postgraduates

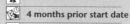 1 : 1

 50%

 Limited

–

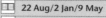 4 months prior start date

3 million volumes

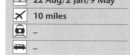 Yes

? Residential

$215 / month

22 Aug/2 Jan/9 May

10 miles

–

–

Utah State University
Biological and Irrigational Engineering Department

UtahState
UNIVERSITY

Contact
Linda John
Biological and Irrigational
Engineering Department
Utah State University
Logan
UT 84322 410
USA

Tel   (+1) 801 797 1181
Fax   (+1) 801 797 1185

Taught Courses
Methods in Biotechnology: cell culture, protein purification techniques; and molecular cloning (three courses); Bioprocesses in engineered and environmental systems; Biochemical engineering; Biomass processing; Land treatment of wastes; Agricultural waste management systems; Principles of irrigation engineering; Drainage engineering for agricultural, urban, and wetland environments; Water supply development and conveyance systems; Sprinkle and trickle irrigation; Surface irrigation design; Irrigation system analysis; Engineering aspects of soil and water conservation; Water management; Principles of remote sensing and applications in agriculture and hydrology; Field irrigation management; Flow measure and control; Optimal groundwater and conjunctive water management I & II; Irrigation system Operations; Drainage investigation and design.

General Departmental Information
The Biological and Irrigation Engineering Department emphasizes a program in agricultural resource engineering addressed toward defining reasonable agricultural goals on a farm, community, regional, continental, or global basis and providing a strategy for organizing and managing water resources with other resources (physical, human, economic, biologic, and natural) that must be brought together to reach desired production and environmental goals.

The department is heavily involved in overseas research and training activities concerned with managing irrigation systems, on-farm water management, and water resources development. The ten professional staff include specializations in: Agricultural hydrology; Bioprocessing; Biotechnology; Crop water-yield analysis; Drainage; evapotranspiration; Food engineering; Groundwater management and simulation; Irrigation conveyance and control structures; Irrigation project planning, design, and operating and management; On-farm water management; Remote sensing and geographical information systems; Surface, sprinkle and trickle irrigation methods.

Research Programs
Research projects in several areas of irrigation and drainage engineering are currently being conducted by the department. Current projects include hydraulics of surface irrigation; Consumptive use; Return flow quantity and quality of irrigation waters and application techniques; Transient flow in tile drainage systems; Drain envelopes; Sprinkler irrigation,; Trickle irrigation; Crop production and water requirements; Salt movement; Regional groundwater modelling for optimizing sustainable yield; Conveyance system modelling and control, and remote sensing.

Specific research projects in the bioprocessing option include ventilation and environmental control of livestock buildings, the contribution of rural municipalities to non-point source pollution, and agricultural waste management systems. Land application of food processing wastes, extrusion of dairy-based foods, multi-stage anaerobic digestion of biological materials, functional properties of foods, and biological detoxification of metals are some of the topics researched in food engineering.

Outstanding Achievement of the Academic Staff to Date
The most recent outstanding staff achievement is that of Dr Lyman Willardson who was recently inducted into the Drainage Hall of Fame. Other notable accomplishments include outstanding ASCE paper in Irrigation and Drainage (Dr Richard Allen) and several staff members have received college research and teacher of the year awards.

Special Departmental Resources and Programs
The Biological and Irrigation Engineering Department is engaged in an extensive program of international irrigation technology transfer, and is contributing significantly to the alleviation of world hunger through multi-lingual training and research in irrigation and drainage. The International Irrigation Center (IIC) was organized to provide an appropriate entity within which to sponsor short-term intensive training activities. The IIC is part of the historic commitment that Utah State university has made for assistance to the developing world. The IIC's objectives are:
– To provide a unique scientifically based practical training experience in agricultural water management and utilization related to the responsibilities and functions at all levels of personnel managing water for social development
– To offer several basic types of training, including short or specialised studies of one to six weeks in the U.S.A. and specific training programs abroad organized by a request from a given country
– Promotion of research involving technology transfer and adaptation of engineering techniques to new environments and sits, with the goal of contributing to the knowledge base and creating research specialist.

📖	$1,206 pa
💲	$3,237 pa
👥	45
👨‍👩‍👧	1 to 4.48
%	50%
📋	Few
👤	Few
🗓	1 January
⬜	Yes
🖥	Yes
?	Small town
🛏	$150 per week
📅	September 1997
✈	83 miles
🚗	–
🚌	83 miles

on the net http://www.editionxii.co.uk

University of Cincinnati
Department of Chemical Engineering

Taught Courses
Core graduate courses in Transport Phenomena, Advanced Thermodynamics, and Chemical Reactor Design. Other graduate courses are taught for a wide range of topics such as: Air Pollution; Aerosol Science; Bioengineering; Ceramic Powders; Powder Technology; Life Science Applications; Membrane Processes; Polymer Systems; Process Dynamics and Control; Process Optimization; Reaction Engineering and Catalysis; Separation Process; Computational Design; Energy and Environmental Science.

Research Programs
Aerosol science; polymers; membranes; catalysis; composite materials; energy and the environment; separation science; process synthesis; and bioseparations.

General Departmental Information
The department offers BS, MS and PhD degrees in Chemical Engineering. It is composed of 14 faculty and currently there are approximately 70 graduate students, almost equally divided between the Masters and Doctoral Programs. All applicants must complete an application form (available on request), submit official copies of all university and college academic transcripts, and arrange for at least two confidential letters of recommendation. Graduate Record Examination (GRE) scores are required. All international applicants whose native language is not English must also submit TOEFL test scores, a score of 580 is the minimum score accepted for admission. Preference will be given to those foreign students who have taken the Test of Spoken English (TSE) and who have scored at least 50 on that test.

Each advanced degree student must satisfy both a teaching and research requirement as part of the graduate training. The teaching requirement is met through assignments as teaching assistants. All full-time graduate students must successfully defend a research-based thesis before graduation. Financial support in the form of stipends and/or tuition awards are available on a competitive basis for highly qualified students.

Outstanding Achievements of the Academic Staff to Date
Please access the department's World Wide Web Site for up to date information.

Special Departmental Resources and Programs
The department research facilities are well equipped to perform research at the highest level. Several of the research laboratories are housed in the newly completed Engineering Research Center, which is dedicated to engineering research and instruction.

Outstanding Local Facilities and Features
Cincinnati is a major manufacturing and financial center, located on the Ohio river in the southwest portion of the state. It has an excellent cultural community including, opera, ballet, symphony, theater, movie theaters. There are outstanding restaurants, museums and a first-class zoo. It also hosts major league sports (baseball and American football). The World Headquarters for Proctor and Gamble are located in the city.

Contact
Professor Robert G Jenkins
Director of Graduate Studies
Department of Chemical
Engineering
PO Box 210171
University of Cincinnati
Cincinnati
OH 45221-0171
USA

Tel (+1) 513 556 2761
Fax (+1) 513 556 3473
Email rjenkins@alpha.che.uc.edu
WWW
http://www.eng.uc.edu/che/

$5,445 (3 Quarters)
$10,383 (3 Quarters)
70
1 : 5
65% (approx)
Yes
–
31 Jan 1997 (fall)
Extensive
Extensive
Urban/self-contained
–
Mid-Sept 1997 (fall)
15 miles (International)
–
–

UNITED STATES

University of Missouri-Rolla
Department of Chemical Engineering

Contact
Partho Neogi
Graduate Student Coordinator
Department of Chemical
Engineering
143 Schrenk Hall
University of Missouri-Rolla
Rolla
MO 65409-1230
USA

Tel   (+1) 573 341 4416
Fax   (+1) 573 341 4377
Email chemengr@umr.edu
WWW
http://www.umr.edu/~chemengr/

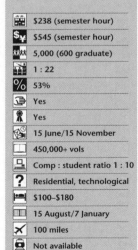

🏛	$238 (semester hour)
💲	$545 (semester hour)
👥	5,000 (600 graduate)
🏫	1 : 22
%	53%
📖	Yes
🧍	Yes
📅	15 June/15 November
📚	450,000+ vols
💻	Comp : student ratio 1 : 10
?	Residential, technological
🛏	$100–$180
📅	15 August/7 January
✈	100 miles
🚗	Not available
🚌	61 miles (Missouri)

Educational Goals of the University of Missouri-Rolla
As Missouri's technological university, the University of Missouri-Rolla's mission is to educate tomorrow's leaders in engineering and science. In a world growing increasingly dependent on science and technology, tomorrow's engineers and scientists must be prepared to be leaders in more than just their chosen professions. They must also be leaders in business, in government, in education, and in all aspects of society. UMR is dedicated to providing leadership opportunities for its students. The opportunity to receive an excellent technological education is only part of the educational experience received by our students. UMR is nationally recognized for its excellent undergraduate and graduate engineering programs, and is distinguished for producing cutting-edge research and key technologies vital to the economic success. UMR has a distinguished faculty dedicated wholeheartedly to the teaching, research, and creative activities necessary for scholarly learning experiences and advancements to the frontiers of knowledge.

Taught Courses
The department offers programs of study leading to MS and PhD programs. A MS non-thesis option is also offered. Areas of study in chemical engineering, which cover both the traditional and nontraditional, range from physical properties measurements to the use of small angle neutron scattering to study protein-protein interactions. These programs cover engineering principles as well as applications at an advanced level and prepare students for careers in such fields as engineering design, operations, research, development, and teaching.

Research Programs
A variety of research areas are available. Specific research areas include: Biotechnology and Environmental Microbiology, Classical and Statistical Thermodynamics, Engineering Data Management, Computer-aided Design and Control, Fluid Mechanics, Transport and Surface Phenomena, Polymers and Separation.

General Departmental Information
You may apply for graduate admission via the World Wide Web. If you apply via the Web you are not required to pay the US $20.00 application fee. You must submit transcripts, three letters of reference, GRE and TOEFL (unless English is native language) scores directly to: Ms Julie Sibley, Admissions Advisor, Admissions Office, University of Missouri-Rolla, 102 Parker Hall, Rolla, MO 65409-1060, USA.

Outstanding Achievements of the Academic Staff to Date
The university was the winner of the 1995 Missouri Quality Award, an award based on the Malcom Baldridge criteria. Ranked in the top 50 engineering schools (1995) by *US News*, ranked #13 by *Money* (1996) as "Best Value in Scientific and Technical Schools."

Faculty text books
C D Holland and A I Liapis, *Computer Methods for Solving Dynamic Separation Problems*, McGraw-Hill Book Company, New York, NY, 1983. This book was translated into Chinese and published in the People's Republic of China in 1988
X J R Avula, R E Kalman, A I Liapis, and E Y Rodin, *Mathematical Modelling in Science and Technology*, Pergamon Press, Elmsford, NY, 1984
A I Liapis, *Fundamentals of Adsorption*, Engineering Foundation, New York, NY (Distributed by the American Institute of Chemical Engineers), 1987
R C Reid, J M Prausnitz, B E Poling, *The Properties of Gases and Liquids*, 4th edition, McGraw-Hill, New York, NY 1987
Rosen, S L, *Fundamental Principles of Polymeric Materials for Practicing Engineers*, Barnes and Nobel, New York, NY 1971
Rosen, S L, *Fundamental Principles of Polymeric Materials*, 2nd edition, Wiley- Interscience, New York, NY 1993
C A Miller and P Neogi, *Interfacial Phenomena: Equilibrium and Dynamic Effects*, in Surfactant Science Series, V 17, Marcel Dekker, Inc, 1986
P Neogi, *Diffusion in Polymers*, Marcel Dekker, Inc, 1996

Fulbright Senior Scholars Abroad
A I Liapis – Germany; D K Ludlow – Israel

Special Departmental Resources and Programs
Students at the University of Missouri-Rolla have access to extensive computing capabilities, including PC and workstation facilities. Many students participate in interdisciplinary projects by interactions with the Intelligent Systems Center, the Graduate Center for Materials Research, the Center for Environmental Science and Technology, the Biochemical Processing Institute, the Rock Mechanics and Explosives Research Center, or the Institute of Thin Film Processing.

New Jersey Institute of Technology

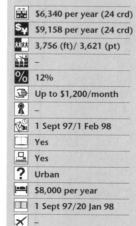

NJIT
New Jersey Institute of Technology

Research Programs

The department's research is supported by approximately $2,5 million annually in grants and contracts from industry and government sources, with principal strengths in the areas of hazardous waste treatment; pharmaceutical processing; membrane separations; and production of solid-state and polymeric materials. NJIT is the lead institute in three environmental research centres involving consortia of universities, industry, and government.

These Centers are housed in a state-of-the-art research facility on the NJIT campus, containing advanced analytical and sensing instruments, as well as bench- and pilot-scale reactors for biological; photochemical; and thermal treatment of hazardous waste.

These Centers, include faculty from the following universities: NJIT; MIT; Ohio State; Penn State; Princeton; Rutgers; Stevens Institute of Technology; Tufts; and the University of Medicine and Dentistry of New Jersey.

The research funding comes from nearly 20 corporate sponsors, US Army; US Environmental Protection Agency (EPA); US Department of Energy; National Science Foundation; and New Jersey Commission on Science and Technology.

In addition, NJIT is a member of a fourth environmental research center headquartered at MIT (and also including the California Institute of Technology), sponsored by the US EPA.

NJIT is also making a major commitment to development of research in solid-state and polymeric materials. Exceptional laboratory facilities exist in polymeric materials (in co-operation with the Polymer Processing Institute, which has more than 40 corporate sponsors); membrane separations (which also enjoys substantial support from industry and government); and solid-state materials (particularly involving chemical vapour deposition reactors).

General Departmental Information

Combining chemical engineering; chemistry; and environmental science in a single department affords a number of interdisciplinary research opportunities within our own faculty.

In addition, we have long enjoyed a close relationship with faculty at the nearby University of Medicine and Dentistry of New Jersey, as well as at the Departments of Biological Sciences and Chemistry of Rutgers University (Newark), which is adjacent to the NJIT campus.

Our programs are also strengthened by close ties to industry.

In addition to refineries and materials processors, New Jersey has the highest concentration of pharmaceutical companies in the United States.

LOCATION

NJIT is located in Newark, NJ, about 20 minutes by public transportation from New York City. It is therefore in the midst of the largest urban complex in the United States, with unequalled availability of cultural activities, from the exalted to the profane.

GRADUATE PROGRAMS – ADMISSIONS AND FINANCIAL SUPPORT

The department offers PhD degrees in Chemical Engineering and in Environmental Science.

The PhD in Chemistry is awarded by the Department of Chemistry at Rutgers University (Newark campus), in which several NJIT faculty hold associated appointments.

The PhD is a research-oriented degree, requiring a minimum of 24 credits of course work (eight courses) beyond the master's degree, and 36 credits of dissertation research. Exceptional students may be accepted directly into the PhD program with a bachelor's degree, and will be required to take an additional 18 credits of course work.

Students must also achieve a satisfactory grade on the doctoral qualifying examination, which tests competence at the master's level, before they can be formally admitted to candidacy.

Admission requirements generally include a minimum grade point average of 3,5 (B+), excellence in the Graduate Record Exam (GRE), and a minimum TOEFL (Test of English as a Foreign Language) score of 550.

Teaching and Research Assistantships are available on a competitive basis, and include (in addition to tuition remission) stipends of approximately $1,200 per month ($14,000 per calendar year).

Contact
Dr Gordon A Lewandowski
Department of Chemical
Engineering, Chemistry and
Environmental Science
New Jersey Institute of
Technology
Newark
NJ 07102
USA

Tel   (+1) 201 596 3573
Fax   (+1) 201 596 8436
Email
lewandows@admin.njit.edu

🏫	$6,340 per year (24 crd)
💲	$9,158 per year (24 crd)
👥	3,756 (ft)/ 3,621 (pt)
🏛	–
%	12%
💲	Up to $1,200/month
👤	–
📅	1 Sept 97/1 Feb 98
☐	Yes
💻	Yes
?	Urban
🛏	$8,000 per year
📑	1 Sept 97/20 Jan 98
✈	–
☎	–
🚌	–

New Jersey Institute of Technology

New Jersey Institute of Technology

Contact
Dr Gordon A Lewandowski
Department of Chemical
Engineering, Chemistry and
Environmental Science
New Jersey Institute of
Technology
Newark
NJ 07102
USA

Tel (+1) 201 596 3573
Fax (+1) 201 596 8436
Email
lewandows@admin.njit.edu

Research Programs

The department's research is supported by approximately $2,5 million annually in grants and contracts from industry and government sources, with principal strengths in the areas of hazardous waste treatment; pharmaceutical processing; membrane separations; and production of solid-state and polymeric materials. NJIT is the lead institute in three environmental research centres involving consortia of universities, industry, and government.

These Centers are housed in a state-of-the-art research facility on the NJIT campus, containing advanced analytical and sensing instruments, as well as bench- and pilot-scale reactors for biological; photochemical; and thermal treatment of hazardous waste.

These Centers, include faculty from the following universities: NJIT; MIT; Ohio State; Penn State; Princeton; Rutgers; Stevens Institute of Technology; Tufts; and the University of Medicine and Dentistry of New Jersey.

The research funding comes from nearly 20 corporate sponsors, US Army; US Environmental Protection Agency (EPA); US Department of Energy; National Science Foundation; and New Jersey Commission on Science and Technology.

In addition, NJIT is a member of a fourth environmental research center headquartered at MIT (and also including the California Institute of Technology), sponsored by the US EPA.

NJIT is also making a major commitment to development of research in solid-state and polymeric materials. Exceptional laboratory facilities exist in polymeric materials (in co-operation with the Polymer Processing Institute, which has more than 40 corporate sponsors); membrane separations (which also enjoys substantial support from industry and government); and solid-state materials (particularly involving chemical vapour deposition reactors).

General Departmental Information

Combining chemical engineering; chemistry; and environmental science in a single department affords a number of interdisciplinary research opportunities within our own faculty.

In addition, we have long enjoyed a close relationship with faculty at the nearby University of Medicine and Dentistry of New Jersey, as well as at the Departments of Biological Sciences and Chemistry of Rutgers University (Newark), which is adjacent to the NJIT campus.

Our programs are also strengthened by close ties to industry.

In addition to refineries and materials processors, New Jersey has the highest concentration of pharmaceutical companies in the United States.

LOCATION

NJIT is located in Newark, NJ, about 20 minutes by public transportation from New York City. It is therefore in the midst of the largest urban complex in the United States, with unequalled availability of cultural activities, from the exalted to the profane.

GRADUATE PROGRAMS – ADMISSIONS AND FINANCIAL SUPPORT

The department offers PhD degrees in Chemical Engineering and in Environmental Science.

The PhD in Chemistry is awarded by the Department of Chemistry at Rutgers University (Newark campus), in which several NJIT faculty hold associated appointments.

The PhD is a research-oriented degree, requiring a minimum of 24 credits of course work (eight courses) beyond the master's degree, and 36 credits of dissertation research. Exceptional students may be accepted directly into the PhD program with a bachelor's degree, and will be required to take an additional 18 credits of course work.

Students must also achieve a satisfactory grade on the doctoral qualifying examination, which tests competence at the master's level, before they can be formally admitted to candidacy.

Admission requirements generally include a minimum grade point average of 3,5 (B+), excellence in the Graduate Record Exam (GRE), and a minimum TOEFL (Test of English as a Foreign Language) score of 550.

Teaching and Research Assistantships are available on a competitive basis, and include (in addition to tuition remission) stipends of approximately $1,200 per month ($14,000 per calendar year).

$6,340 per year (24 crd)	
$9,158 per year (24 crd)	
3,756 (ft)/3,621 (pt)	
–	
12%	
Up to $1,200/month	
–	
1 Sept 97/1 Feb 98	
Yes	
Yes	
Urban	
$8,000 per year	
1 Sept 97/20 Jan 98	
–	
–	
–	

Rensselaer Polytechnic Institute
School of Engineering

Taught Courses

AERONAUTICAL ENGINEERING, MEng, MS, DEng, PhD

BIOMEDICAL ENGINEERING, MEng, MS, DEng, PhD

CHEMICAL ENGINEERING, MEng, MS, DEng, PhD

CIVIL ENGINEERING, MEng, MS, DEng, PhD

COMPUTER AND SYSTEMS ENGINEERING, MEng, MS, DEng, PhD

DECISION SCIENCES AND ENGINEERING SYSTEMS, PhD

ELECTRICAL ENGINEERING, MEng, MS, DEng, PhD

ELECTRIC POWER ENGINEERING, MEng, MS, DEng, PhD

ENGINEERING PHYSICS, MS, PhD

ENGINEERING SCIENCE, MS, PhD

ENVIRONMENTAL ENGINEERING, MEng, MS, DEng, PhD

INDUSTRIAL AND MANAGEMENT ENGINEERING, MEng, MS

MANUFACTURING SYSTEMS ENGINEERING, MS

MATERIALS SCIENCE AND ENGINEERING, MEng, MS, DEng, PhD IN MATERIAL ENGINEERING

MECHANICAL ENGINEERING AND MECHANICS, MEng, MS, DEng, PhD

NUCLEAR ENGINEERING, MEng, MS, DEng, NUCLEAR ENGINEERING AND SCIENCE, PhD

OPERATIONS RESEARCH AND STATISTICS, MS

TRANSPORTATION ENGINEERING, MEng, MS, DEng, PhD

General Departmental Information
The faculty is characterized by achievers and doers. Of the approximately 135 tenure-track faculty, 98% already hold their PhDs and in 1995–96, they were awarded over $41 million in sponsored research.

Rensselaers's optimum size blends individual program flexibility with broad based research opportunities, Rensselaer is known for its leadership and innovation in engineering education. US News & World Report ranks our graduate engineering programs in the first tier (top 25) overall, and in the top 10 in reputation by practicing engineers.

Outstanding Achievements of the Academic Staff to Date
Many well known publications: *Fundamentals of Physics*, by Robert Resnick, Halliday and Resnick, *Elementary Differential Equations and Boundary Value Problems* by W E Boyce and R C DiPrima, *Behavior in Organizations* by Robert Baron, *Oriental Philosophies*, by John Koller, and from Schaum's Outline Series, *Theory and Problems of Thermodynamics*, by Michael Abbott and Hendrick Van Ness.

Special Department Resources and Programs
Center for Integrated Electronics and Electronic Manufacturing (CIEEM); Center for Multiphase Research (CMR); Center for Services Sector Research and Education (CSSRE); Scientific Computation Research Center (SCOREC); Center for Composite Materials and Structures (CCMS); Lighting Research Center; Fresh Water Institute (FWI); Center for Image Processing Research (CIPR); Center for Infrastructure and Transportation Studies (CITS); New York State Center for Advanced Technology in Automation; Robotics; and Manufacturing (CAT); New York State Center for Polymer Synthesis (CPS); and Center for Entrepreneurship of New Technological Ventures (CENTV).

Contact
Vicki Lynn
Assistant Dean of Engineering
Rensselaer Polytechnic Institute
School of Engineering
Room 3004
Troy
New York 12188-3590
USA

Tel (+1) 518 276 6203
Fax (+1) 518 276 8788
Email lynnv@rpi.edu
WWW
http://www.rpi.edu/dept/grad~admissions

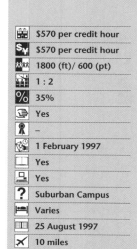

$570 per credit hour
$570 per credit hour
1800 (ft)/ 600 (pt)
1 : 2
35%
Yes
–
1 February 1997
Yes
Yes
Suburban Campus
Varies
25 August 1997
10 miles
10 miles
10 miles

UIC

University of Illinois at Chicago
Civil and Materials Engineering

Contact
Mohsen A Issa
University of Illinois at Chicago
Department of Civil and
Materials Engineering
2095 Engineering Research
Facility, M/C 246
842 West Taylor Street
Chicago
Illinois 60607 7023
USA

Tel   (+1) 312 996 3432
Fax   (+1) 312 996 2426
Email MIssa@uic.edu.

Taught Courses
MS AND PhD IN CIVIL ENGINEERING AND MS AND PhD IN MATERIALS ENGINEERING
Courses are taught in: Structural Mechanics and Dynamics; Design and Behaviour of Reinforced and Prestressed Concrete and Steel Structures; Transportation Engineering and Planning; Water Resources and Environmental Engineering; Geotechnical and Geoenvironmental Engineering; Mechanics of Solids; Materials Processing and Characterization; Materials Engineering.

Research Programs
Research includes: Structural Engineering and Mechanics: static and dynamic analysis of linear and non-linear structures, including modal analysis, failure process of concrete structures, bridge rehabilitation and rating, structural dynamics and seismic response of structures, concrete fracture, damage and creep, and microstructure of materials; Environmental Engineering: surface and groundwater hydrology and hydraulics, evaporation and other interactions between land surface and the atmosphere, groundwater contamination and transport, geoscience aspects of waste disposal, and management of stored and spilled hazardous waste; Geotechinal Engineering: environmental geotechnics, earthquake engineering, pavement analysis, and geomechanics; Transportation Engineering: surface transportation facilities (infrastructure), planning, operations and maintenance through computer-based methods including knowledge-based system, and design and evaluation of intelligent transportation systems; Materials Engineering: physical and mechanical characterization of various materials, fatigue and fracture, corrosion and welding, as well as processing.

General Department Information
The Department of Civil and Materials Engineering faculty aims at establishing a strong orientation towards research as the basic foundation of graduate education. An extensive array of research is carried out in the department. The department is well facilitated with equipment that is state-of-the-art in terms of technology and materials, including universal testing machines, acoustic emission systems, X-ray diffraction systems, and a wide variety of microscanning systems. The department is located in a newly constructed building, and has recently acquired several laboratories. The computer facilities available include a network of 25 workstations which are equipped with the latest software and hardware to perform civil engineering design and analysis related tasks as well as the necessary word processing, graphics, and computation capabilities. The library system is very comprehensive, and is connected to other libraries within the State of Illinois. The CME Department has its own machine shop that employs two full-time machinists with a combined experience of over 30 years. They are responsible along with two full-time equipment specialists for the fabrication of special experimental test setups, operation of newly acquired equipment, technical support, etc.

Special Departmental Resources and Programs
The Department has specialized laboratories in the following research areas:

The Structural and Concrete Materials Laboratory includes several state of the art equipment used for controlled testing of various structural components and materials in a closed-loop system. The laboratory also includes acoustic emission systems, concrete analysis systems, and high speed computer controlled data acquisition systems.

The Geotechnical and Geoenvironmental Engineering Laboratory has complete capabilities for both field and laboratory investigations to determine strength and hydraulic properties of geomaterials and to investigate foundation systems during landslides, earthquakes, and flooding. The laboratory is well developed in investigating ways to use waste and recycled materials in civil engineering, innovative soil and groundwater remediation methods, and designing effective solid and hazardous waste landfills.

The Fracture Research Facility is designed for conducting detailed failure analysis and prediction that includes the development of fracture mechanism maps for advanced materials such as: polymers and their composites, ceramics and ceramic composites, carbon-carbon composites, and metal–non-metal laminates.

Outstanding Local Facilities and Features
Museum of Science and Industry, Field Museum of Natural History, Chicago Art Institute and Museum of Modern Art, John G. Shedd Aquarium, Alder Planetarium, Brookfield Zoo, World Music Theatre and Extensive Live Theatre, Chicago Jazz and Blues Festival, Chicago Symphony Orchestra, Lyric Opera, Home of the World Champion Chicago Bulls (basketball), White Sox and Cubs (baseball), Blackhawks (hockey), and Bears (football).

Date for Final Applications

	Spring 1998	Summer 1997	Fall 1997
Internationals:	July 26, 1997	January 3, 1997	March 21, 1997
US Nationals:	October 18, 1997	April 4, 1997	May 30, 1997
Classes begin:	January 13, 1998	June 2, 1997	August 25, 1997

🏛	$2,349/semester
💲	$5,272/semester
👥	24,589
🏢	–
%	–
🍽	Yes
🧑	–
🛏	See main text
🖥	–
💻	–
?	–
🛏	$3,000/semester
📅	August 97 and January 98
✈	16 km
🚊	800 m
🚌	400 m

Utah State University
Civil and Environmental Engineering Department

Taught Courses
Finite elements; Limit analysis of structures; Structural stability; Similitude numerical methods; Prestressed concrete; Structural dynamics; Structural reliability; Composite structures; Geographic information systems; Pavement design; Transportation systems analysis; Traffic engineering; Geometric design transportation; Earth and rock fill dams; Deep foundations; Environmental geotechnics; Soil mechanics; Earthquake engineering; Physical hydrology; Surface runoff hydrology; Engineering risk and reliability; Groundwater hydroclimatology; Groundwater modelling; Subsurface contaminant transport; Hydraulic modelling; Open channel flow; Sedimentation engineering; Potential fluid flow; Air quality management; Solid and hazardous waste; Water and wastewater treatment; Industrial wastewaters; Water resource engineering; others.

Research Programs
Steel bridge design; Concrete; GIS; Transportation; Earthquake; Soils; Wastewater; Industrial engineering; Natural systems; Groundwater; Hydrology; Water resources; Hydraulics; Fluid mechanics.

General Departmental Information
The Civil and Environmental Engineering Department at Utah State University has approximately 190 graduate students.

The Department has five graduate divisions: Structures; Geotechnical; Transportation; Environmental; and Water.

Our programs are well known nationally and internationally. Our faculty consists of 35 members.

Outstanding Achievements of the Academic Staff to Date
Several of our faculty members are involved in international research as well as research in the United States. Some of countries that work has been done in are: Australia; Thailand; India; Senegal; Mauritania; Mali; Morocco; and Southern American countries.

Contact
Becky Hansen
Civil and Environmental
Engineering Department
Utah State University
Logan
UT 84322 410
USA

Tel (+1) 801 797 2938
Fax (+1) 801 797 1185

	$1,206 pa
	$3,237 pa
	15,118
	1 : 4,48
%	50%
	Few
	Few
	1 January
	Yes
	Yes
?	Small town
	$150 per week
	September 1997
	83 miles
	–
	83 miles

Oregon Graduate Institute of Science and Technology
Department of Electrical Engineering

Contact
Margaret Day
Registrar
Oregon Graduate Institute of
Science and Technology
20000 NW Walker Road
PO Box 91000
Portland
Oregon 97291-1000
USA

Tel   (+1) 800 685 2423
  or (+1) 503 690 1028
Fax (+1) 503 690 1686
Email
admissions@admin.ogi.edu
  or msee@ee.ogi.edu
WWW   http://www.ogi.edu

Taught Courses
Graduate level courses (MS & PhD) offered in the following areas of specialization

COMMUNICATIONS AND SIGNAL PROCESSING

INFORMATION PROCESSING

ELECTRONIC CIRCUIT DESIGN

SEMICONDUCTOR PROCESSING, DEVICE PHYSICS AND MATERIALS

VLSI PROCESSING AND DESIGN

MULTIMEDIA

COMPUTATIONAL FINANCE

BIOMEDICAL ENGINEERING

Research Programs
Research programs leading to a Masters Thesis and/or PhD degree are conducted in many of the academic specialization areas. The department publishes a research overview describing faculty background and current research programs, available upon request.

General Departmental Information
Oregon Graduate Institute is centrally located amid many high technology companies (Intel, Tektronix, Sequent, Hewlett-Packard, SEH, Triquent). Due to the strong demand for electrical engineering professionals with advanced technical training, our department provides a quality accredited academic program leading to a Master of Science degree valued highly by high-tech industries. Since electrical engineering has become both diverse and interdisciplinary, students with BA/BS degrees in math, physics, computer science, engineering, or related fields are encouraged to apply.

Outstanding Local Facilities and Features:
OGI is located in the "silicon forest" of the northwestern United States, near Portland, Oregon. Portland offers multicultural, social and recreational diversity in art, music, food, theater, sports, and retail. Also associated with its location, OGI is only a short drive away from outdoor recreational areas offering hiking, camping, skiing, sailboarding, as well as scenic mountain ranges and the spectacular Oregon coast.

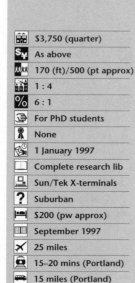

🏫	$3,750 (quarter)
💲	As above
👥	170 (ft)/500 (pt approx)
🏢	1 : 4
%	6 : 1
📑	For PhD students
🧍	None
🕐	1 January 1997
📖	Complete research lib
💻	Sun/Tek X-terminals
?	Suburban
🛏	$200 (pw approx)
📅	September 1997
✈	25 miles
🚉	15–20 mins (Portland)
🚌	15 miles (Portland)

Rensselaer Polytechnic Institute
School of Engineering

Taught Courses

AERONAUTICAL ENGINEERING, MENG, MS, DENG, PHD

BIOMEDICAL ENGINEERING, MENG, MS, DENG, PHD

CHEMICAL ENGINEERING, MENG, MS, DENG, PHD

CIVIL ENGINEERING, MENG, MS, DENG, PHD

COMPUTER AND SYSTEMS ENGINEERING, MENG, MS, DENG, PHD

DECISION SCIENCES AND ENGINEERING SYSTEMS, PHD

ELECTRICAL ENGINEERING, MENG, MS, DENG, PHD

ELECTRIC POWER ENGINEERING, MENG, MS, DENG, PHD

ENGINEERING PHYSICS, MS, PHD

ENGINEERING SCIENCE, MS, PHD

ENVIRONMENTAL ENGINEERING, MENG, MS, DENG, PHD

INDUSTRIAL AND MANAGEMENT ENGINEERING, MENG, MS

MANUFACTURING SYSTEMS ENGINEERING, MS

MATERIALS SCIENCE AND ENGINEERING, MENG, MS, DENG, PHD IN MATERIAL ENGINEERING

MECHANICAL ENGINEERING AND MECHANICS, MENG, MS, DENG, PHD

NUCLEAR ENGINEERING, MENG, MS, DENG, NUCLEAR ENGINEERING AND SCIENCE, PHD

OPERATIONS RESEARCH AND STATISTICS, MS

TRANSPORTATION ENGINEERING, MENG, MS, DENG, PHD

General Departmental Information

The faculty is characterized by achievers and doers. Of the approximately 135 tenure-track faculty, 98% already hold their PhDs and in 1995–96, they were awarded over $41 million in sponsored research.

Rensselaers's optimum size blends individual program flexibility with broad based research opportunities, Rensselaer is known for its leadership and innovation in engineering education. US News & World Report ranks our graduate engineering programs in the first tier (top 25) overall, and in the top 10 in reputation by practicing engineers.

Outstanding Achievements of the Academic Staff to Date

Many well known publications: *Fundamentals of Physics*, by Robert Resnick, Halliday and Resnick, *Elementary Differential Equations and Boundary Value Problems* by W E Boyce and R C DiPrima, *Behavior in Organizations* by Robert Baron, *Oriental Philosophies*, by John Koller, and from Schaum's Outline Series, *Theory and Problems of Thermodynamics*, by Michael Abbott and Hendrick Van Ness.

Special Department Resources and Programs

Center for Integrated Electronics and Electronic Manufacturing (CIEEM); Center for Multiphase Research (CMR); Center for Services Sector Research and Education (CSSRE); Scientific Computation Research Center (SCOREC); Center for Composite Materials and Structures (CCMS); Lighting Research Center; Fresh Water Institute (FWI); Center for Image Processing Research (CIPR); Center for Infrastructure and Transportation Studies (CITS); New York State Center for Advanced Technology in Automation; Robotics; and Manufacturing (CAT); New York State Center for Polymer Synthesis (CPS); and Center for Entrepreneurship of New Technological Ventures (CENTV).

Contact
Vicki Lynn
Assistant Dean of Engineering
Rensselaer Polytechnic Institute
School of Engineering
Room 3004
Troy
New York 12188-3590
USA

Tel   (+1) 518 276 6203
Fax   (+1) 518 276 8788
Email lynnv@rpi.edu
WWW
http://www.rpi.edu/dept/
grad~admissions

$570 per credit hour
$570 per credit hour
1800 (ft)/ 600 (pt)
1 : 2
35%
Yes
–
1 February 1997
Yes
Yes
Suburban Campus
Varies
25 August 1997
10 miles
10 miles
10 miles

UNITED STATES

University of Florida
Electrical and Computer Engineering

Contact
Dr Leon Couch
Assoc. Chairman & Professor
PO Box 116200
University of Florida
Gainesville
FL 32611
USA

Tel   (+1) 352 392 4931
Fax   (+1) 352 392 8671
WWW
http://www.ece.ufl.edu

Taught Courses

The ECE Department offers graduate programs leading to the degrees of:

– **MASTER OF ENGINEERING**

– **MASTER OF SCIENCE; ENGINEER**

– **DOCTOR OF PHILOSOPHY**
The nine areas of major concentration are:

– **COMMUNICATIONS**

– **COMPUTER ENGINEERING**

– **DEVICE AND PHYSICAL ELECTRONICS**

– **DIGITAL SIGNAL PROCESSING**

– **ELECTRIC ENERGY SYSTEMS**

– **ELECTROMAGNETICS**

– **ELECTRONIC CIRCUITS**

– **PHOTONICS**

– **SYSTEMS AND CONTROL**
Course work and research opportunities are offered in many specializations. These include: control systems; digital communications; digital hardware and signal processing; electric energy and power electronics; electronic noise; image processing and computer graphics; integrated and fiber optics; laser electronics and fabrication; lighting; microprocessors; radar; reliability of semiconductor devices; robotics; signal estimation; semiconductor devices; wireless communications; spread spectrum systems; telecommunication systems; VLSI circuit design; fabrication and characterization.

Teaching and Research Facilities

The ECE Department occupies 85,000 square feet of space with a large portion devoted to funded research. The research facilities include IC processing in a Class 100 environment with 0.7 micron capability. Computing facilities are excellent including more than seventy mini-computers and workstations, hundreds of PCs, and extensive industrial software packages.

General Information

The University of Florida is located in Gainesville (190,000 inhabitants in the metropolitan area). The UF enrolment is approximately 40,000 with over 140 departments and professional colleges in Law; Dentistry; Medicine; and Veterinary Medicine.
The ECE Department has an enrolment of 450 undergraduate students and 300 graduate students.

APPLYING

For application forms contact the:
Graduate Studies Office
Department of Electrical and Computer Engineering
227 Larsen Hall
PO Box 116200
University of Florida
Gainesville
FL 32611
USA
Email: gsbro@admin.ee.ufl.edu
Professor Gijs Bosman, ECE Graduate Co-ordinator, may be contacted at gbosm@admin.ee.ufl.edu.

	$115/credit hour
	$386/credit hour
	40,000
	1 : 9
	20%
	Yes
	Yes
	7 June 1997
	Yes
	Yes
	2,000 acres, Urban
	US $250/month
	Late August 1997
	6 miles
	24 hours
	800 miles

Utah State University
Electrical and Computer Engineering Department

Taught Courses
Analog electronics; Communications; Controls; Digital systems; E and M; Electro-Optics; Microwaves; Signals and systems; Space engineering; VLSI.

Research Programs
DSP; Communications; VLSI; Space science; Control systems; Parallel processing; Instrumentation; Electroacoustics; Image and signal processing; Antennas and RF radios; Pattern recognition; Molecular design.

General Departmental Information
The ECE Department offers the following graduate degrees: Master of science; Master of engineering; Electrical engineer; and Doctor of philosophy.

Admission decisions are based on undergraduate curriculum and GPA, GRE and TOEFL scores; letters of recommendation; and previous work experience.

New graduate students are normally admitted for fall quarter only, since many courses are taught in sequence.

Decisions are made early in the calendar year for the following fall.

Outstanding Achievements of the Academic Staff to Date
Several of our faculty have received the college's outstanding teacher and outstanding faculty advisor awards. The majority of our faculty are also involved heavily in research, and some have received national and international recognition for achievements and advancements in their respective areas.

Many work very closely with leading companies in conjunction with their research, giving students working with them hands-on experiences in industry.

During the past five years, three members of our faculty received outstanding researcher awards and another was chosen as outstanding alumnus by Princeton University.

Special Departmental Resources and Programs
Students are exposed to real-world problem solving from research projects. Both students and faculty receive funding to support their efforts in those projects. Contacts with industry and governmental agencies also provide opportunities for internships and permanent job placement. Joint publications with faculty and students enhance knowledge dissemination and opportunity for student growth and involvement.

Outstanding Local Facilities and Features
Festival of the American West is held annually for a week during the summer and includes a pageant, displays, and exhibits.

The Ellen Eccles Theatre; Lyric Theatre; and various smaller local theatre companies present outstanding musical and dramatic productions year-round.

USU has a good museum where displays are changed periodically. A variety of concerts and speakers are also sponsored by the university regularly.

Contact
Robert Gunderson
Electrical and Computer
Engineering Department
Utah State University
Logan
UT 84322 410
USA

Tel   (+1) 801 797 2938
Fax   (+1) 801 797 1185

$1,206 pa	
$3,237 pa	
15,118	
1 to 4,48	
50%	
Few	
Few	
1 January	
Yes	
Yes	
Small town	
$150 per week	
September 1997	
83 miles	
–	
83 miles	

UNITED STATES

NDSU

North Dakota State University
College of Engineering & Architecture

Contact
Waneta Truesdell
College of Engineering
and Architecture
North Dakota State University
1400 Centennial Boulevard
PO Box 5285
Fargo
North Dakota 58105 5285
USA

Tel   (+1) 701 231 7494
Fax   (+1) 701 231 7195
Email
truesdel@badlands.nodak.edu
WWW
http://www.ndsu.edu/

Taught Courses
The College of Engineering & Architecture has departments in:
– Agricultural and Biosystems Engineering
– Architecture
– Landscape Architecture
– Aero Manufacturing Technology
– Civil Engineering
– Construction Management
– Electrical Engineering
– Industrial Engineering and Management
– Mechanical Engineering

Research Programs
Air and Water Pollution: Automated Controls; Bioengineering Communication; Control Systems, Electrical & Mechanical; Electromagnetics, Electronic; Engineering Mechanics, Thermal Systems, Environmental engineering; Food Engineering; Geotechnical and Soils Physics; Hazardous Waste; Hydrology and Biological Systems; Machine Design; Manufacturing Information Systems; Manufacturing Processes; Materials, Ferrous and Non-Ferrous Power Systems; Signal Processing Computer; Solid Waste Management; Structural Engineering; Water Resources.

General Departmental Information
The college has the latest computer facilities and many lecturers and well equipped labs. North Dakota State University is the land grant university of North Dakota. The College has a student enrollment of 2300 and a faculty of 82. It has an outstanding reputation having supplied CEO's and Vice Presidents to some of the leading companies in the USA, such as Boeing, 3M, etc. Many of its alumni have formed their own successful corporations. Besides these a number of alumni are engaged with consulting firms and in governmental agencies. To this end, graduates from NDSU are highly sought after.

Outstanding Achievements of the Academic Staff to Date
In the past year, the faculty have produced:
– One book sent to printers
– 43 journal articles
– 46 conference proceedings papers, and numerous circulars, bulletins and monographs.
Typical achievements have been made in previous years. For further details, please refer to our WWW page: http://www.ndsu.edu/

Special Departmental Resources and Programs
Graduate courses are available in most departments. Programs of study may be arranged leading to the Master of Science degree in the engineering fields and to the Master of Architecture degree. The Doctor of Philosophy in Engineering degree is a single doctoral program in the College of Engineering & Architecture administered by the Graduate School and the College of Engineering and Architecture. The PhD program is characterized by an interdisciplinary approach to engineering. A single, doctoral program for agricultural engineering, civil engineering, electrical engineering, industrial engineering, and mechanical engineering provides studies with both general knowledge and in-depth understanding of one major area of concentration.

Fargo was rated as the fifth best best city in the United states concerning quality of life by Money magazine. While winter temperatures do reach sub-zero, summers are pleasant and the people are good-hearted and friendly. Other qualities contributing to the reputation for desirable living include: the highly respected educational system, advanced medical technology, a progressive business community, numerous cultural and arts opportunities, a variety of stores and shops and clean air and water. With more than 150,000 people in the area, Fargo-Moorhead is the largest metropolitan center between Minneapolis and Seattle. Located on the Red River, the border between North Dakota and Minnesota, Fargo-Moorhead is rich in fertile farmlands and has three universities and the business and technical colleges providing education for more than 23,000 students.

	$1,160 per smtr (12 cr)
	$3,097 per smtr (12 cr)
	9,700
	1 : 20
%	10%
	Yes
	Yes
	March 1997
	Yes
	Yes
?	Land Grant
	$98
	Fall Semester
	2 miles
	2 days
	1,141 miles

Purdue University
Schools of Engineering

Programs of Study

Purdue University offers one of the most comprehensive and highly-ranked engineering programs in the United States. The Schools of Engineering at Purdue offer MS and PhD degrees in each of the following nine disciplines: Aeronautics and Astronautics, Agricultural and Biological Engineering, Chemical Engineering, Civil Engineering, Electrical and Computer Engineering, Industrial Engineering, Materials Engineering, Mechanical Engineering, and Nuclear Engineering. Interdisciplinary programs, such as Biomedical Engineering and Operations Research, are offered under the auspices of one or more of the schools; other interdisciplinary programs may be arranged.

Research Programs

The principal facilities for the Schools of Engineering are housed in ten buildings on the campus and five research laboratories on or near the campus. Each school maintains its own laboratories, as well as computer facilities and specialized equipment for research. Students also have access to the Engineering Computer Network, the University Computer Center, and the University and Engineering libraries.

The Institute for Interdisciplinary Engineering Studies, located in the A A Potter Engineering Center, was established for interdisciplinary research activities in such areas as energy, biomedical, and transportation engineering. The Engineering Research Center for Collaborative Manufacturing is a cross-disciplinary unit also located in the Potter Center. Annual expenditures for research in all of the Engineering programs exceeds 75 million dollars.

General Information

The Schools of Engineering have approximately 8,200 students on the main campus, of whom 1,800 are enrolled as graduate students. The total student population on the campus is 36,000.

The West Lafayette campus, which has its own airport, is situated near the banks of the Wabash River on 1,565 acres. Tippecanoe County and the surrounding areas offer a variety of cultural activities as well as historic landmarks and recreational attractions. The campus is located 60 miles from the state capital at Indianapolis and 130 miles from Chicago.

Contact
Dr Warren H Stevenson
Assistant Dean of Engineering
Purdue University
1280 Engineering
Administration Building
West Lafayette
IN 47907-1280
USA

Tel   (+1) 317 494 5340
Fax   (+1) 317 494 9321
Email
stevensw@ecn.purdue.edu

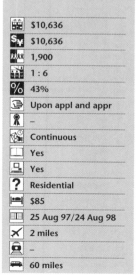

$10,636
$10,636
1,900
1 : 6
43%
Upon appl and appr
–
Continuous
Yes
Yes
Residential
$85
25 Aug 97/24 Aug 98
2 miles
–
60 miles

University of Southern California
School of Engineering

Contact
Margery Berti
Assistant Dean of Engineering
University of Southern California
School of Engineering
Olin Hall 33OG
Los Angeles
CA 90089-1454
USA

Tel   (+1) 213 740 6241
Fax   (+1) 213 740 8493
Email berti@mizar.usc.edu

Taught Courses
The School of Engineering offers graduate programs leading to the MS, Engineer, and PhD degrees in: Aerospace Engineering; Applied Mechanics; Biomedical Engineering; Chemical Engineering; Civil Engineering; Computer Engineering; Computer Science; Electrical Engineering; Engineering Management; Environmental Engineering; Industrial and Systems Engineering; Manufacturing Engineering; Materials Engineering; Materials Science; Mechanical Engineering; Operations Research; petroleum engineering, and systems architecture and engineering. The Master of Construction Management is also offered. Areas of emphasis such as biomedical imaging and telemedicine; computer networks; multimedia and creative technologies; robotics and automation; software engineering; and VSLI design are available within certain majors.

Requirements for the master's degree are satisfied by completion of a minimum of 27 units of graduate course work; a thesis may be written as part of this requirement. The Engineer degree requires 30 units of course work beyond the MS degree, with concentration in two areas of engineering, and a qualifying examination. The PhD requires 60 units beyond the bachelor's degree and a qualifying examination. Students must demonstrate an ability to make an original contribution to a chosen field through independent study and research. A minimum of one year of research is required after passing the qualifying examination. The student must defend the dissertation before a suitably constituted committee. A minimum grade point average of 3.0 must be earned in all graduate programs.

Research Programs
The School of Engineering occupies 11 buildings on the west side of the campus at University Park in Los Angeles and has extensive research facilities. The Engineering and Sciences Library in the Seaver Science Center has extensive collections in all areas of engineering. Research institutes operating under the School's auspices include Biomedical Simulations Resource; Center for research on Applied Signal Processing; Centre Research on Environmental Sciences, Policy and Engineering; Center for Software Engineering; Communication Sciences Institute; Electronics Science Laboratory – The Joint Services Electronics Program; Information Sciences Institute; Center for Manufacturing and Automation Research; Institute for Robotics and Intelligent Systems; Integrated Media Systems Center; Los Angeles Rubber Group National; Center for Integrated Photonic Technology; Research Centre for Computational Geomechanics; Signal and Image Processing Institute; Southern California Center for Advanced Transportation Technologies; Western Research Application Center.

Computer facilities for instruction and research are provided by the University Computing Services. The Student Computing Facility, dedicated to instruction, houses over 100 Sun SPARCstation, UNIX workstations, and 25 X-terminals in public user rooms with access to Sun SPARCstation 20 file servers and timesharing systems. The Research Computing Facility supports general purpose, parallel, and statistical computing and is equipped with Sun SPARCserver 600s, a SPARCcenter 1000, an IBM SP2, a Convex SPP-1000, and an SGI Power Challenge. High-speed access to a CRAY C98/8128 at the San Diego Supercomputer Center is also available.

Special Departmental Resources and Programs
A unique feature of the USC School of Engineering's graduate programs is the customised master's degrees. Several departments offer students the possibility of choosing an area of specialization within the degree program. Some examples are the MS in Biomedical Engineering with the specialization in biomedical imaging and telemedicine; the MS in Computer Science with a specialization in computer networks or robotics and automation or software engineering; and the MS in Electrical Engineering with the specialization in computer networks or VSLI design. In response to the intense interest in multimedia, two new specializations have just been added in multimedia and creative technologies for students earning the MS in Computer Science or the MS in Electrical Engineering.

General Departmental Information
Full-time students typically take 9–12 units per semester. Mandatory fees are $251 per semester. Teaching/research assistantships and fellowships are available. Various University meal plans can be purchased for approximately $1,460 per semester.

Privately owned off-campus apartments can be rented in the immediate vicinity of the campus for $500–$800 per month. Housing is available in the Greater Los Angeles areas at all levels of costs.

Outstanding Achievements of the Academic Staff to Date
14 members (more than 10%) of the active faculty and four emeritus faculty have been elected to the National Academy of Engineering, which is one of the highest professional distinctions accorded to an engineer. 20 faculty have received the coveted National Young Investigator Award from the National Science Foundation. Two of these have also received the prestigious and highly competitive Presidential Faculty Fellows Award given by the National Science Foundation in recognition of the scholarly achievements and potential of the US' most outstanding young science and engineering faculty. Over 40 faculty are fellows of engineering societies.

The USC School of Engineering has been ranked second nationally in sponsored research expenditures and research volume per faculty member by *US News* and *World Report* every year since 1990.

Outstanding Local Facilities and Features
Only five minutes by freeway from the Center of Los Angeles, USC is secluded on an extensive, landscaped campus with a quiet academic atmosphere. Students may take advantage of the broad cultural offerings of a major metropolis. The Los Angeles County Museum of Art, the Los Angeles Philharmonic Orchestra, and the Los Angeles Music Center are all close by. Entertainment options of every description are available within a short driving distance.

	$6,100 per semester (95–96)
	$6,100 per semester (95–96)
	–
	1 : 14
%	1,000+ international students
	Yes
	No
	Fall, Spring and Summer
	Yes
	Yes
?	Residential
	$1,700 (1995–96)
	25 Aug/5 Jan/10 May
	15 miles
	4.5 hours
	400 miles

on the net http://www.editionxii.co.uk

University of Tennessee, Knoxville
College of Engineering

UNITED STATES

Taught Courses
Graduate programs in the College of Engineering at The University of Tennessee, Knoxville provide opportunities for advanced study leading to the Master of Science and the Doctor of Philosophy degrees in the fields of aerospace; agricultural; chemical; civil, electrical; mechanical; metallurgical; nuclear; and polymer engineering and engineering science.

Additionally, the MS degree is offered in environmental engineering and industrial engineering.

Research Programs
The MS and PhD degree programs listed above reside in eight academic departments which make up the College of Engineering. Each department is responsible for conducting a robust, comprehensive research program. Much of the research is interdisciplinary in nature. Examples of ongoing research activities in the various degree programs are listed below:

Aerospace Engineering – aerodynamics; aircraft design, performance, and control; helicopter rotor dynamics; space re-entry.

Agricultural Engineering – bioprocessing; food engineering; machinery systems; sensors, instrumentation, and automatic controls; soil and water engineering; waste management.

Chemical Engineering – process control; molecular modelling; waste product minimisation, recovery, and utilisation; bioprocesses; rheology and processing of polymers.

Civil Engineering – transportation systems; construction materials; structural engineering; geotechnical engineering; public works engineering.

Electrical Engineering – communications; instrumentation and controls; electronics; high voltage and insulation; image processing; motors and drives; pattern recognition; industrial plasma engineering; power electronics; robotics; systems and networks.

Engineering Science – computational fluid dynamics; computational solid mechanics; biomedical engineering; mechanics of composite materials; dynamics and vibrations; fracture mechanics.

Environmental Engineering – air quality; water quality; waste management; remediation and reclamation; water resources, risk assessment.

Industrial Engineering – manufacturing and production systems; human factors, ergonomics, and biomechanics; operation simulation and optimisation.

Mechanical Engineering – alternate fuels; vehicle emission controls; robotics; machine design and product development; fluidized bed technologies; materials casting and thermal control; heat transfer and thermodynamics; thermoeconomics.

Metallurgical Engineering – welding and joining; high-temperature materials; surface treatments; corrosion behaviour; physical metallurgy; materials characterisation.

Nuclear Engineering – reactor physics; reactor instrumentation and control; radiation protection engineering; reactor safety; shielding; fusion; risk assessment; radioactive waste management; advanced maintenance and reliability engineering.

Polymer Engineering – polymer processing; polymer crystallisation; composite materials; polymer morphology; rheology; mechanical, physical, and chemical behaviour of polymers

Engineering faculty also participate in three major Centres of Excellence including the Science Alliance, the Centre for Materials Processing, and the Waste Management Research and Education Institute. Some of these centres are collaborative with other colleges and research organisations such as Oak Ridge National Laboratories.

The College also participates in the Transportation Centre, the Measurement and Controls Engineering Centre, and the Maintenance and Reliability Centre. All of these centres involve heavy participation by industrial partners. The theoretical and applied work in computational mechanics is conducted in a state-of-the-art Centre for Computer Integrated Engineering.

General Information
The College of Engineering is one of twelve colleges which make up The University of Tennessee, Knoxville (UTK), a comprehensive research university founded in 1794. Currently, the total enrolment at UTK is about 26 000 students, approximately 7 000 of whom are pursing graduate degrees in more than 85 fields. Approximately 800 students are currently pursuing MS and PhD degrees in the twelve graduate majors available within the College of Engineering. The work of these graduate students is coordinated by 135 full-time tenure-track faculty members plus many adjunct and research faculty members.

The 215-hectare campus of UTK is located in Knoxville, a city of 300 000 in eastern Tennessee. The university is located on the banks of the Tennessee River in the foothills of the beautiful Great Smoky Mountains.

APPLYING
Initial inquiry and application should be made to The Graduate School. A minimum B average is required. For students whose native language is not English, a minimum TOEFL score of 550 is required. GMAT scores are not required, but GRE scores are required in most programs.

Contact
Graduate Admissions
and Records
218 Student Services Building
The University of Tennessee
Knoxville
Tennessee 37996 0220
USA

Tel (+1) 423 974 3251
Fax (+1) 423 974 6541
Email gsinfo@utk.edu
WWW
http://www.utk.edu

Associate Dean for
Academic Services
College of Engineering
101 Perkins Hall
Knoxville
Tennessee 37996 2011
USA

Tel (+1) 423 974 2454
Email fgilliam@utk.edu

🏠	$1,372 per semester
💲	$3,540 per semester
👥	6,603
🏫	1:5
%	15%
📖	Yes
🧍	–
🗓	1 March for Fall admission
🖥	Yes
🖥	Yes
?	Urban, public
🛏	$102 for meals and lodging
🗓	Fall 1997: 27 August
✈	10 miles (15 km)
🚗	–
🚆	–

Contact
Dr Gordon A Lewandowski
Department of Chemical
Engineering, Chemistry and
Environmental Science
New Jersey Institute of
Technology
Newark
NJ 07102
USA

Tel   (+1) 201 596 3573
Fax   (+1) 201 596 8436
Email
lewandows@admin.njit.edu

New Jersey Institute of Technology

Research Programs
The department's research is supported by approximately $2,5 million annually in grants and contracts from industry and government sources, with principal strengths in the areas of hazardous waste treatment; pharmaceutical processing; membrane separations; and production of solid-state and polymeric materials. NJIT is the lead institute in three environmental research centres involving consortia of universities, industry, and government.

These Centers are housed in a state-of-the-art research facility on the NJIT campus, containing advanced analytical and sensing instruments, as well as bench- and pilot-scale reactors for biological; photochemical; and thermal treatment of hazardous waste.

These Centers, include faculty from the following universities: NJIT; MIT; Ohio State; Penn State; Princeton; Rutgers; Stevens Institute of Technology; Tufts; and the University of Medicine and Dentistry of New Jersey.

The research funding comes from nearly 20 corporate sponsors, US Army; US Environmental Protection Agency (EPA); US Department of Energy; National Science Foundation; and New Jersey Commission on Science and Technology.

In addition, NJIT is a member of a fourth environmental research center headquartered at MIT (and also including the California Institute of Technology), sponsored by the US EPA.

NJIT is also making a major commitment to development of research in solid-state and polymeric materials. Exceptional laboratory facilities exist in polymeric materials (in co-operation with the Polymer Processing Institute, which has more than 40 corporate sponsors); membrane separations (which also enjoys substantial support from industry and government); and solid-state materials (particularly involving chemical vapour deposition reactors).

General Departmental Information
Combining chemical engineering; chemistry; and environmental science in a single department affords a number of interdisciplinary research opportunities within our own faculty.

In addition, we have long enjoyed a close relationship with faculty at the nearby University of Medicine and Dentistry of New Jersey, as well as at the Departments of Biological Sciences and Chemistry of Rutgers University (Newark), which is adjacent to the NJIT campus.

Our programs are also strengthened by close ties to industry.

In addition to refineries and materials processors, New Jersey has the highest concentration of pharmaceutical companies in the United States.

LOCATION
NJIT is located in Newark, NJ, about 20 minutes by public transportation from New York City. It is therefore in the midst of the largest urban complex in the United States, with unequalled availability of cultural activities, from the exalted to the profane.

GRADUATE PROGRAMS – ADMISSIONS AND FINANCIAL SUPPORT
The department offers PhD degrees in Chemical Engineering and in Environmental Science.

The PhD in Chemistry is awarded by the Department of Chemistry at Rutgers University (Newark campus), in which several NJIT faculty hold associated appointments.

The PhD is a research-oriented degree, requiring a minimum of 24 credits of course work (eight courses) beyond the master's degree, and 36 credits of dissertation research. Exceptional students may be accepted directly into the PhD program with a bachelor's degree, and will be required to take an additional 18 credits of course work.

Students must also achieve a satisfactory grade on the doctoral qualifying examination, which tests competence at the master's level, before they can be formally admitted to candidacy.

Admission requirements generally include a minimum grade point average of 3,5 (B+), excellence in the Graduate Record Exam (GRE), and a minimum TOEFL (Test of English as a Foreign Language) score of 550.

Teaching and Research Assistantships are available on a competitive basis, and include (in addition to tuition remission) stipends of approximately $1,200 per month ($14,000 per calendar year).

$6,340 per year (24 crd)	
$9,158 per year (24 crd)	
3,756 (ft)/3,621 (pt)	
–	
12%	
Up to $1,200/month	
–	
1 Sept 97/1 Feb 98	
Yes	
Yes	
Urban	
$8,000 per year	
1 Sept 97/20 Jan 98	
–	
–	
–	

on the net   http://www.editionxii.co.uk

Alfred University
College of Engineering and Professional Studies/New York State
College of Ceramics/School of Ceramic Engineering and Sciences

Alfred University

Programs of Study
Programs that lead to master's and doctoral degrees are open to those with baccalaureate degrees in ceramic science and engineering, chemical engineering, chemistry, electrical engineering, geology, glass science, industrial engineering, materials science, mechanical engineering, metallurgy, and physics.

Master of Science degrees are awarded in ceramic engineering, ceramic science, and glass science. The research track requires 15 credit hours of course work and 15 credit hours of an experimental thesis. The manufacturing option requires 24 credit hours, including at least six hours in business and management, and a six-credit dissertation, which must be defended orally. Master of Science degrees in electrical, industrial, and mechanical engineering require 24 credit hours of course work and six credit hours of thesis. An oral presentation is required.

The Doctor of Philosophy degree is offered in ceramics and glass science. Ninety hours of course credit beyond the requirements for the baccalaureate degree must be earned. A minimum of 40 credit hours must be in regular course work, with the remainder as thesis credits.

Research Facilities
In addition to a complete stock of routine scientific equipment, the following major items and facilities are available: transmission and scanning electron microscopes; electronic microprobe; X-ray diffraction equipment; X-ray fluorescence and IR analysis equipment; a laser-Raman spectrograph; a secondary-ion mass spectrometer; Instron Universal Tester for mechanical properties; glassblowing, machine, and electronics facilities with technical personnel; and chemical analysis and microscopic-sample preparation facilities. The College library has extensive holdings in ceramics, materials science, electrical engineering, industrial engineering, and mechanical engineering and in relevant areas of chemistry, physics, and mathematics.

Financial Aid
Assistantships sponsored by the college and fellowships supported by private industry and government agencies are available. Assistantships include an annual stipend of $12,675, health insurance, and tuition remission. Postdoctoral fellowships are available.

Cost of Study
Tuition and fees for ceramics and glass engineering and science are $11,940 (publicly suported), and $18,972 for electrical, industrial, and mechanical engineering for 1996–97. Laboratory fees are $500 per semester.

Living and Housing Costs
The estimated room and board charge is $6,002 or less a year.

Student Group
Alfred University has 1,831 undergraduate and 400 graduate students from 43 states and ten countries. There are ninety-two graduate students in the School of Ceramic Engineering and Sciences and the Divisions of Electrical, Industrial, and Mechanical Engineering.

Student Outcomes
Graduates are in demand, with all of PhD graduates finding employment as research scientists at such companies as Corning Inc and Battelle PNL. MS graduates found employment as research scientists and engineers at companies such as with places like Corning Inc; Buffalo China; Owens Corning; AVX Corp; Libbey-Owens Ford; and Norton Company.

Location
Alfred University is located in Alfred, NY, which is 70 miles south of Rochester, NY; 90 miles southeast of Buffalo, NY; and 60 miles west of Corning, NY. It is six hours' drive from New York City.

Applying
Candidates for admission to the Graduate School must demonstrate the ability to perform at the graduate level. The non-refundable application fee is $50.

Contact
Katherine McCarthy
Director of Admissions
Suite A
Alfred University
Alfred
New York 14802-1232
USA

Tel   (+1) 607 871 2141
WWW
http://www.nyscc.alfred.edu
or
http://www.alfred.edu

University of Arkansas
College of Engineering

Contact
Dr Robert C Welch
University of Arkansas
College of Engineering
4188 Bell Engineering Center
Fayetteville
AR 72701
USA

Tel (+1) 501 575 6010
Fax (+1) 501 575 4346
Email RCW@engr.uark.edu
WWW
http://www.engr.uark.edu/

Taught Courses
Postgraduate courses are taught in the following engineering disciplines:

BIOLOGICAL AND AGRICULTURAL ENGINEERING	ENVIRONMENTAL ENGINEERING
CHEMICAL ENGINEERING	INDUSTRIAL ENGINEERING
CIVIL ENGINEERING	MECHANICAL ENGINEERING
COMPUTER SYSTEMS ENGINEERING	OPERATIONS MANAGEMENT
ELECTRICAL ENGINEERING	OPERATIONS RESEARCH
ENGINEERING MANAGEMENT	TRANSPORTATION ENGINEERING

Distance education courses leading to a Master of Science in Engineering degree are offered worldwide. These courses may also be used to meet course work requirements for the PhD degree.

SUBJECT AREAS OF ENGINEERING RESEARCH
Advanced Manufacturing Technology; Amorphous Silicon; Applications of Ag Chemicals; Chemical Spills; Computer Architecture; Control Systems; Cost Engineering; Deposit, Thin Film; Electric Machines; Electronic Packaging; Electronics Manufacturing; Environmental Monitoring; Fiber Composites Reinforcement; Food Processing; Food Safety; Health Care Systems; High Performance Concrete; High Temperature Superconductors; Human Factors; IC Design; IC Modelling; Image Processing; Image Understanding; Imaging Medical; Manufacturing Systems; Manufacturing Technology; Materials Management; Mixing; Modelling Aids; Multichip Modules; Operations Research; Optical Character Recognition; Pattern Recognition; Pavement Design; Pavement Management; Petroleum Dewaxing; Poultry Housing; Power System Quality; Process Modelling; Radar Imaging; Radiation Calibration and Dosimetry; Reliability; Robotics; Separations, BIO; Separations, Membrane; Silicon Microprobes; Simulation; Soil Erosion; Solar Cells; Solid Waste; Synthetic Diamond; Traffic Signing; Transportation Logistics; Transportation Planning; Wastewater Treatment; Water Management; Water Quality; Wet Lands.

General Information
The 100 members of the engineering faculty offer seven undergraduate degree programs, 12 master's degree programs and the PhD. The library holdings of 1 500 000 volumes and those of AMIGOS and OCLC are assessible through the Internet. More than 600 PCs and 110 Sun workstations are networked and available for students use. Laboratories and research centers are equipped with state-of-the-art equipment for students use. Distance education courses use the TV production facilities of the college.

Outstanding Achievements of the Academic Staff to Date
Over 6 000 books and papers have been published by the engineering faculty.

Special Departmental Resources and Programs
The college operates the following specialised centers:
– Graduate Engineering Distance Education Center
– Arkansas Center for Technology Transfer
– Chemical Hazards Research Center
– Genesis Technology Business Incubator
– High-Density Electronics Center
– Logistics Center
– Mack-Blackwell Rural Transportation Study Center
– Southwest Radiation Calibration Center
– Systems Technology Center

Outstanding Local Facilities and Features
The University of Arkansas is located in Fayetteville, a progressive college town with a population of 50 000. Situated in the Ozark Mountains of Northwest Arkansas, the region is one of the fastest growing areas in the United States and ranked as one of the best areas to live in the country.

🏫	$780/three hour course
💲	$1,350/three hour course
👥	14,902
👨‍🏫	17
%	8%
📖	Yes
🧍	Yes
📅	30 days prior to semester
📖	Yes
💻	Yes
?	Coed (Residential)
🛏	$115–$150
📅	26 Aug/13 Jan/19 May
✈	3 miles
🚗	–
🚌	192 miles

Clemson University
College of Engineering and Science

Taught Courses

Clemson University's College of Engineering and Science offers graduate degree programs in 20 areas. The following areas have graduate programs and degrees: Agricultural Engineering (MEngr, MS, PhD); Bioengineering (MS, PhD); Ceramic Engineering (MEngr, MS, PhD); Chemical Engineering (MEngr, MS, PhD); Chemistry (MS, PhD); Civil Engineering (MEngr, MS, PhD); Computer Engineering (MS, PhD); Computer Science (MS, PhD); Electrical Engineering (MEngr, MS, PhD); Engineering Mechanics (MS, PhD); Environmental Systems Engineering (MEngr, MS, PhD); Hydrogeology (MS); Industrial Engineering (MS, PhD); Materials Science & Engineering (MS, PhD); Mathematical Sciences (MS, PhD); Mechanical Engineering (MEngr, MS, PhD); Physics (MS, PhD); Textile and Polymer Science (PhD); Textile Chemistry (MS, PhD); Textile Science (MS).

Research Programs

Agricultural and Biological Engineering: Aquaculture, agricultural-waste management, biotechnology, crop production, mechanization, animal housing, residential housing, global climate change and air quality, soil and water management, crop processing, instrumentation and controls, food and process engineering; **Bioengineering**: Biomaterials, biomechanics, bioinstrumentation; **Ceramic Engineering**: Traditional ceramics, ceramic fibers, ceramic matrix composites, glass, electronic ceramics, sol-gel processing, aerogels, hazardous waste vitrification and management; **Chemical Engineering**: Thermodynamics, polymers, process automation, kinetics, separations; **Chemistry**: Bioorganic chemistry, flourine chemistry, photochemistry, polymer chemistry, organometallic chemistry, environmental sampling and analysis, mass spectrometry, chemical physics; **Civil Engineering**: Applied fluid mechanics, construction materials, projects management, structural and geotechnical engineering; **Computer Engineering**: Computer communications, computer systems architecture, communications/ digital signal processing, controls/robotics; **Computer Science**: Computer architecture, database management systems, design and analysis of algorithms, operating systems, parallel computation, programming languages, graphical systems, networks, systems measurements and modeling, software engineering, theory of computing; **Electrical Engineering**: Computational electromagnetics, communications/digital signal processing, computer communications; control/robotics, electronics, power systems; **Engineering Mechanics**: Mechanics of composite materials and structures; **Environmental Systems Engineering**: Process engineering, hazardous and radioactive waste treatment, contaminant characterization, contaminant fate and transport, analysis of natural systems, environmental restoration, risk assessment and waste management; **Hydrogeology**: Groundwater geology, subsurface remediation; **Industrial Engineering**: Integrated manufacturing, quality engineering, human factors engineering, manufacturing systems, operations research; **Materials Science and Engineering**: Application of fundamental principles that govern the structure of the solid state to produce optimum mechanical, electrical, optical, and other physical properties; **Mathematical Sciences**: Algebra/combinatorics, analysis, computational mathematics, operations research, statistics/probability; **Mechanical Engineering**: Applied mechanics, mechanical and manufacturing systems, thermal/fluid science; **Physics and Astronomy**: Astrophysics, atmospheric physics, biophysics, computational physics, radiation physics, experimental and theoretical solid-state physics, theoretical physics; **Textile and Polymer Science**: Fiber physics, textile structures, fiber formation, polymer science; **Textile Chemistry**: Fiber chemistry, polymer chemistry, chemistry of dyeing and/or finishing fibers and textiles, chemistry of composite systems; **Textile Science**: Fiber science, polymer science, textile technology.

General Departmental Information

Research is an integral part of graduate study in engineering and science, with research laboratories located in each of the 16 buildings on the Clemson campus. Among facilities of special interest are a ceramic pilot lab, molecular simulation facilities, a microelectronics clean-room facility, wind engineering lab, 15,000-gpm recirculating flume, computer-aided manufacturing facilities, machine shops, an optical flow measurements lab, a gas turbine flow facility, and a pressure casting facility. There are also a variety of labs for the study of Experimental animal surgery and histopathology; Artificial intelligence; Microchemistry analysis; Experimental thermodynamics; Computer process control; Polymer synthesis, characterization, and processing; Polymeric resorbable biomaterials; polymer composites; Radioactive materials; Hazardous and toxic materials analysis; Bioorganic chemistry; Flourine chemistry; Mass spectrometry; Microelectronics; Image processing; Speech processing; Fluid mechanics; Electrooptics and lasers; Antennas and microwaves; Semiconductor reliability; High-speed cinematography and stroboscopy; Heat transfer; Solar cell testing and reliability; Materials processing; Robotics; Aerodynamics; Metallurgy; Electrical fabrication, repair, and calibration; Atmospheric phenomena through the use of radar, rocket, and satellite data; Production of man-made and natural fibers; Coloring agents and finishes; Chemistry of composite systems; and textile technology.

Contact
Christian E G Przirembel
Associate Dean of Engineering and Science for Research and Graduate Studies
College of Engineering and Science
114 Riggs Hall
Box 340901
Clemson University
Clemson
South Carolina 29634-0901
USA

Tel   (+1) 864 656 3200
Fax   (+1) 864 656 0859
Email
chris.przirembel@ces.clemson.edu
WWW
http://www.ces.clemson.edu

🏫	$1,461 / semester
💲	$2,922 / semester
👥	920 graduate students
🏢	1 : 3
%	–
📖	Yes / Assistantships
👤	–
📅	15 June 1997 (for Fall)
⬜	Yes
💻	Yes
?	Public 4 year
🛏	–
📅	August 97/January 98
✈	40 mins
🚗	N/A
🚌	130 miles

UNITED STATES

The University of Dayton

University of Dayton
Graduate Engineering Program

Contact
Donald L Moon
University of Dayton
300 College Park
Dayton
OH 45469-0227
USA

Tel   (+1) 513 229 2241
Fax   (+1) 513 229 2471
Email
dmoon@engr.udayton.edu
WWW
http//www.engr.udayton.edu

Taught Courses
The School of Engineering offers several program leading to the graduate degrees of Doctor of Philosophy in Engineering (Aerospace, Electrical, Materials, & Mechanical); Doctor of Philosophy in Electro optics; Doctor of Engineering (Aerospace, Electrical, Materials, & Mechanical); Master of Science in Aerospace Engineering; Master of Science in Chemical Engineering; Master of Science in Civil Engineering; Master of Science in Electrical Engineering; Master of Science in Electro Optics; Master of Science in Engineering Management; Master of Science in Engineering Mechanics; Master of Science in Materials Engineering; Master of Science in Mechanical Engineering; Master of Science in Engineering; Master of Science in Management.

Research Programs
Signal Processing (Statistical Signal Processing, Optical Signal Processing, Image Processing, Wavelets), Electro Optical Systems and Devices (Systems Design, System Characterization, Performance Improvement). Electromagnetic (Radar, microwave, integrated waveguides), Reliability, Expert Systems, Structures, Micromechanics of Composites, Hazardous waste engineering Wastewater treatment, Water distribution systems, geotechnical and earthquake engineering, carbon materials, electrochemical processes, high temperature materials, liquid solid and gas liquid agitation, polymers, mechanical behaviour (fatigue and fracture, corrosion, non-destructive evaluation), ceramics (structural, superconductivity) composites (mechanics, processing), modern control and robotics, electronics (microelectronics, VHDL), mechanics and design.

General Departmental Information
The University of Dayton, a privately endowed institution, was established in 1850. The School of Engineering was started in 1910, and graduate engineering activities were initiated in 1961. The University annually engages in approximately $50 million of engineering and scientific research through the sponsorship of private industry and government agencies. The School of Engineering has 55 full-time faculty members, approximately 70 graduate assistants and a full complement of staff personnel.

Special Departmental Resources and Programs
The modern Eugene W. Kettering Engineering and Research Laboratories building, a six-story, air conditioned facility, provides 211,000 square feet and contains 88 laboratories, 14 classrooms, 115 faculty offices, and eight seminar rooms. Practically all of the engineering technology, and research activities are housed in this building. Laboratories are available in such areas as man-maching simulation, energy conversion, electronics, electro optics, digital systems, magnetics, systems and human performance, microwaves, holographics, manufacturing, metallurgy, are plasma, high temperature materials, materials analysis, structural mechanics, civil engineering, spectroscopy, environmental engineering, soil mechanics, and aerospace, instrumentation, and fluid mechanics. The campus computer facilities are directly connected with all the engineering and research activities. Many computers are available for faculty and student use. Research programs are conducted in the basic engineering and natural science areas, as well as in interdisciplinary activities involving psychology, behavioural science, and biophysics.

Outstanding Local Facilities and Features
Greater Dayton is a metropolitan area of about 933,500 people, situated on the Miami River in south-western Ohio. It is approximately an hour's drive from Metropolitan Cincinnati and Columbus. As a result, a variety of cultural and recreational are available. The Dayton area is also in a favourable academic climate, with fifteen colleges and universities seven of which offer graduate work located within 75 miles. The Dayton area has one of the highest ratios, of engineers and scientists to total population in the nation, and there is a heavy concentration of industrial and government engineering activity.

🏛	$379/$427 credit hour
💲	$379/$427 credit hour
👥	500
🎓	1 : 20
%	10.7%
📖	Yes
🧑	Yes
📅	1 August 1997
🏠	Yes
💻	Yes
?	Urban, private
🛏	$125–$150
📅	20 Aug/4 Jan/10 May
✈	10 miles
🚉	45 minutes
🚌	70 miles

University of Maryland Baltimore County
College of Engineering

Taught Courses
The University of Maryland Baltimore County (UMBC) College of Engineering offers undergraduate and graduate programs in Chemical Engineering, Computer Sciences, and Mechanical Engineering as well as a graduate program in Electrical Engineering. A joint program leading to a MS in Engineering Management is offered by UMBC and the University of Maryland's University College (UMUC). This program is designed for people working full-time in engineering positions.

Research Programs
Bioengineering; Biomechanics; Communications; Downstream Bioprocessing; Fluid Mechanics; Heat Transfer; Manufacturing; Photonics; Robotics; Signal Processing; Upstream Bioprocessing; Intelligent Information Systems; High-Performance Computing and Communications; Computer Graphics; Algorithms and Theory.

General Departmental Information
The College of Engineering is currently funded for $5.2 million in research grants. Research in the Chemical and Biochemical Engineering Department is heavily focused in biochemical engineering and covers a wide range of areas. Mechanical Engineering speciality areas include biomechanics design, manufacturing and system mechanics. The Computer Science and Electrical Engineering Department provides advanced instruction, training, and research opportunities to prepare students for careers and marketable skills in business, industry, universities and government agencies.

Outstanding Achievements of the Academic Staff to Date
UMBC College of Engineering faculty members have received many prestigious awards. Among these are four National Science Foundation Professional Young Investigator Awards and two National Science Foundation Career Development Awards.

Special Departmental Resources and Programs
The Chemical Engineering Department possesses $2 million in equipment for carrying out biochemical engineering. The Mechanical Engineering Department has labs for materials testing, heat transfer, biomaterials, robotics and mechatronics research. The Computer Science and Electrical Engineering Department has substantial computing resources acquired with the new (1992) Engineering and Computer Science Building including the Onyx Reality Engine2, SGI Indigo workstations, SGI Crimson and SPARC servers. SUN workstations and a network of Apple Macintosh and 486/Pentium-based personal computers.

Outstanding Local Facilities and Features
Baltimore is home to several theaters, art museums, concert halls, the Inner Harbor, annual ethnic festivals, and professional football and baseball teams. It is centrally located, with easy access to New York City (four hours), Philadelphia (one hour), and Washington DC (45 minutes). Countless cultural activities are readily accessible.

Contact
Dr Shlomo Carmi
Dean
College of Engineering
University of Maryland
Baltimore County
1000 Hilltop Circle
Baltimore
Maryland 21250
USA

Tel   (+1) 410 455 3270
Fax   (+1) 410 455 3559

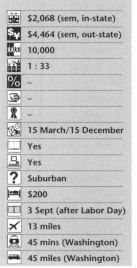

🏛	$2,068 (sem, in-state)
💲	$4,464 (sem, out-state)
👥	10,000
🏫	1 : 33
%	–
🏦	–
🧍	–
⏰	15 March/15 December
🛏	Yes
🖥	Yes
?	Suburban
🛌	$200
📅	3 Sept (after Labor Day)
✈	13 miles
🚉	45 mins (Washington)
🚌	45 miles (Washington)

UNITED STATES

University of Miami
School of Engineering

Contact
Dr Thomas D Waite
University of Miami
College of Engineering
1251 Memorial Drive
McArthur Room 243
Coral Gables
FL 33146
USA

Tel   (+1) 305 284 3467
Fax (+1) 305 284 2885
Email twaite@eng.miami.edu

Taught Courses
College of Engineering graduate courses

MS PROGRAM
Architectural; biomedical; civil; electrical and computer; industrial.

PhD PROGRAM
Biomedical; civil; electrical and computer; industrial; mechanical; industrial ergonomics.

DA PROGRAM
Civil; mechanical.

Dual MS degree – offered through joint program with the School of Business Administration. DA degree – designed to prepare students for college teaching. Offered with cooperation of the School of Education. Master's degree programs available with/without a thesis in most areas of specialization.

Requirements
MS program with thesis
Approved integrated program – minimum of 30 semester credits; average grade B or better; at least nine course credits on 600 level; approved thesis of six credits; oral examination in defence of thesis.

MS program without thesis
Approved integrated program; minimum of 36 semester credits; average grade of B or better; at least 12 course credits on 600 level; oral examination after completion of minimum 18 credits of graduate work.

PhD program
Candidates are expected to complete appropriate integrated program of studies in preparation for the comprehensive qualifying examination; preparation normally requires two years after bachelor's degree; one to two years beyond qualifying examination will usually be needed for completion of acceptable dissertation and remaining course work; student then admitted to final oral examination.

DA program
78 credits beyond baccalaureate are required; at least 48 credits must be in major or cognate fields; nine credits in higher education; 12 credits in internship and a project.

Research Programs
In the area of *Biomedical Engineering* the following basic areas are treated: Biiomedical instrumentation and artificial organs, applications of computers to diagnostic and information systems, signal and image processing, engineering principles applied to physiological system analysis, biomechanics, biofluid dynamics, tissue engineering, clinical engineering, optics and lasers, and rehabilitation engineering. In *Civil and Architectural Engineering* the areas of study include: structures, structural mechanics, environmental engineering, water resources engineering, engineering management, ocean engineering and biomedical engineering with integrated programs in the supporting areas of: Architecture, mathematics and computer science, geological sciences, chemistry, biology, oceanography, physics, radiology, mechanical engineering, industrial engineering, computer science and business administration. Concentrations are also offered in: building systems/engineering and environmental systems. In *Electrical and Computer Engineering* options include: communications, computers and electronics, engineering management, ocean engineering, biomedical engineering, medical informatics, computer architecture and digital systems, computer software, electronics, signal processing, machine intelligence, and communications. In *Industrial Engineering* the areas of concentration include: ergonomics, health care systems, engineering management, manufacturing engineering, occupational health and safety, productivity engineering, management of technology, operations research and quality. Finally, *Mechanical Engineering* offers: fluid mechanics, heat transfer, energy conversion, hydrogen energy, environmental engineering, HVAC systems, materials science, solid mechanics, robotics, controls and design, as well as Engineering Management.

General Departmental Information
Students in the university number 13,541 and come from 49 states and over 100 countries. Students of all religions, races, and nationalities participate in the objective of the university, which is to produce graduates who can make significant contributions to the society in which we live. Founded in 1925, the university is nonprofit, nondenominational, and coeducational. It is free from religious and political control and derives its funds from tuition and from direct contributions from alumni and others interested in its work in teaching and research.

Special Departmental Resources and Programs
Computer-Aided Engineering Applications Laboratory; Laboratories for Pollution Control; Edward Arnold Multi-Media Center; Computer Vision and Image Processing; Device Simulation Laboratory; Digital and Analog Filter Laboratory; Electronic Design Automation/Computer-Aided Software Engineering Laboratory; Underwater Imagining Laboratory; VLSI Design Laboratory; Integrated Computer Manufacturing; Biomechanics Research; Work Physiology and Ergonomic Research; Human Factors and Aging Research; Productivity Research and Industrial Hygiene; Clean Energy Research Institute; Fluids and Thermal Sciences Laboratory; Internal Combustion Engines Laboratory; Remote Sensing Laboratory; HVAC&R and Two-Phase Flow Laboratory; Dorgan Solar Energy Laboratory; Computational Fluid Mechanics Laboratory; Bio-fluidmechanics; Tissue Mechanics; Medical Imaging; Neurological and Audiological Instrumentation; Rehabilitation Engineering; Lasers and Optics in Medicine.

	$742 (credit)
	$742 (credit)
	177 (masters/PhD)
	1 : 10
	32%
	Yes
	Yes
	No specific date)
	Otto Richter Library
	5 labs (public)
	Subtropical/coastal/urban
	–
	29 Aug/13 Jan
	–
	–
	–

on the net   **http://www.editionxii.co.uk**

The University of Michigan
College of Engineering

Taught Courses
There are 11 departments comprising the teaching foci of the College of Engineering

ATMOSPHERIC, OCEANIC AND SPACE SCIENCES

AEROSPACE ENGINEERING

BIOMEDICAL ENGINEERING

CHEMICAL ENGINEERING

CIVIL AND ENVIRONMENTAL ENGINEERING

ELECTRICAL ENGINEERING AND COMPUTER SCIENCE

INDUSTRIAL AND OPERATIONS ENGINEERING

MATERIALS SCIENCE AND ENGINEERING

MECHANICAL ENGINEERING AND APPLIED MECHANICS

NAVAL ARCHITECTURE AND MARINE ENGINEERING

NUCLEAR ENGINEERING AND RADIOLOGICAL SCIENCES

Research Programs
There are over 150 engineering-related research centers and laboratories affiliated with the College of Engineering at the University of Michigan. The following major laboratories and centers illustrate the breadth of research being undertaken at the College: Automated Semiconductor Manufacturing Center; Automotive Research Center; Center for Ergonomics; Cooperative Institute for Limnology and Ecosystems Research; Dimensional Measurement and Control in Manufacturing Center; Display Technology and Manufacturing Center; Electron Microbeam Analysis Laboratory; Institute for Environmental Science, Engineering and Technology; Intelligent Transportation Systems Research Center of Excellence; Laboratory for Scientific Computation; Macromolecular Science and Engineering Center; Michigan Memorial Phoenix Project; Michigan Sea Grant College Program; Reconfigurable Machining Engineering Research Center; Space Physics Research Laboratory; Center for Utrafast Optical Sciences; Tauber Manufacturing Institute; Virtual Reality Laboratory.

General Departmental Information
The University of Michigan College of Engineering, ranked among the top engineering schools of the world, was founded in 1853–54, when fewer than a half-dozen other American colleges were providing formal study in engineering. It established the first programs in the US in Metallurgical Engineering, Naval Architecture and Marine Engineering, Electrical Engineering, Chemical Engineering, Aeronautical Engineering, Nuclear Engineering and Computer Engineering. Today, approximately 2,000 graduate students and approximately 300 faculty members engage in teaching, learning and research in state of the art facilities totalling over 1.5 million square feet (166,000 square meters). In 1995–96, the college awarded over 650 graduate degrees, and expended over $95 million US dollars in research.

Outstanding Achievements of the Academic Staff to Date
Many of the faculty are fellows in disciplinary societies and are members of prestigious national committees which set the agenda for education and research in the US. Also among the faculty are 10 National Engineering Academy members.

Special Departmental Resources and Programs
Media Union
The Media Union is a 225,000 square-foot (25,000 square-meter) integrated technology instruction center that houses collections of information resources normally found in a traditional library, and also provides high-tech equipment to further explore the physical and simulated world. Some of the activities envisioned for the Media Union include: Exploring new avenues for creative collaboration; Extending human cognition and sensory powers; Inventing new paradigms for learning and teaching; Discovering new art forms; Designing new buildings for new forms of community; Creating new devices for industry and home.

Joint graduate programs
The College of Engineering has a strong commitment to interdisciplinary study, offering several graduate programs administered jointly with other nationally-ranked schools and colleges such as the School of Business, the College of Architecture and Urban Planning, the School of Public Health, the School of Music, and others.

Outstanding Local Facilities and Features
The city of Ann Arbor, located in the scenic Huron River Valley, has a population of approximately 130,000 and is known as a friendly, picturesque town with big city amenities. Local, national and international musicians, dance troupes, and theatre companies frequent the town's theatres and concert halls. The town contains science and fine art galleries with internationally renown collections, and hosts several nationally recognized art fairs. Major collegiate sporting events occur throughout the year in premier sporting facilities. Other cultural and entertainment activities, including professional sporting venues are available only minutes away, in the greater Detroit area.

Contact
Graduate Program Chair
Department of (specify)
The University of Michigan
College of Engineering,
Administration
Robert H Lurie Engineering
Center
1221 Beal Avenue
Ann Arbor
Michigan 48109-2102
USA

Tel (+1) 313 647 7000
WWW
http://www.engin.umich.edu/

$10,786	
$20,352	
1,918	
1 : 6.5	
42%	
Through depts	
N/A	
15 Jan–1 Feb 1997	
23 libs, 6.6 mil vols	
45 comp sites	
Residential	
$342 (int)	
First week Sept 1997	
35 miles (57.4 km)	
16 hours 30 mins	
514 miles (828 km)	

University of Mississippi
Engineering School

Contact
Dr Allie M Smith
School of Engineering
101 Carrier Hall
The University of Mississippi
at Oxford
University
MS 38677
United States

Tel   (+1) 601 232 7407
Fax   (+1) 601 232 1287
Email enas@olemiss.edu
WWW
http://www.olemiss.edu/depts/
engineering_school/

Taught Courses
Graduate Studies can be carried out in Engineering Science leading to MS and PhD degrees with specific emphasis in the following areas: Chemical; Civil; Electrical; Geological; Environmental and Mechanical Engineering; Computer Science and Engineering; Geology; Materials Science and Engineering; Computational Hydroscience and Engineering, and Telecommunications.

There are also MS and PhD degree programs in Computational Engineering Science.

Research Programs
In Chemical Engineering: Research is being conducted in Coal combustion; Biochemical engineering; Stochastic modelling; Soil remediation; Process control; and Integrated circuit fabrication.
In Civil Engineering: Research is being conducted in Highway pavement design and management; Finite element modelling; Structural dynamics and seismic analysis; Mechanics of composite materials; Computational simulation; and Acoustics.
In Computer Science: Research is being conducted in Software systems; Internet technology software engineering and development; Machine learning; Computer graphics and simulation; and Computer engineering.
In Electrical Engineering: Research is being conducted in Electromagnetic fields and waves; Computational electromagnetics; Non-destructive testing; Remote sensing; Wireless communications; and Antennas.
In Engineering Science: Research is being conducted in Wireless communications; Computational engineering science; Environmental engineering; Hydraulics; and Energy conservation.
In Geology and Geological Engineering: Research is being conducted in Minerals; Remote sensing; Geographic information systems; Design of waste disposal repositories; Waste management and recycling; Groundwater contaminant transport; and Geologic mapping.
In Mechanical Engineering: Research is being conducted in Computational hydroscience and fluid mechanics; Heat transfer and thermophysics; Materials science; Composite materials and pultrusion; Experimental fluid mechanics; and Finite element analysis.
In Telecommunications: Research is being conducted in Two-way messaging; Personal communications systems; Wireless local networks; Very short range applications of wireless communications technology; and Computer-to-computer communications.
In Computational Hydroscience and Engineering: Research is being conducted in Environmental flows; Sedimentation processes; Water resources engineering and computational modelling and animation of flows; Soil erosion; and Sediment transport.
In Mineral Resources: Research is being conducted in Environmental/geotechnical mapping; Earthquake hazard identification; Ground water geochemistry; Marine geophysical systems development; Marine mineral resource evaluations; and Development of acoustical systems for environmental applications and downhole well maintenance.

General Departmental Information
The School of Engineering, which began engineering instruction in 1854, is part of The University of Mississippi, which has over 10,000 students on its Oxford campus of which 2,250 are graduate students. The campus is extremely attractive as one would expect for a classical resident Southern USA University which is currently celebrating the 150 year anniversary of its founding.

Outstanding Achievements of the Academic Staff to Date
Faculty are nationally and internationally recognized. They have over 700 scholarly research publications in the last five years, have chaired numerous scholarly conferences, are editors of numerous books and archival journals, and are recipients of numerous national research awards. Several are Fellows of ASCE; ASME; IEEB; and AIAA.

Special Departmental Resources and Programs
Graduate students in the School of Engineering have tremendous access to very extensive super-computing capabilities. In composite materials processing, the School has one of only two pultrusion machines on a college campus in the USA. A state-of-the-art scanning electron microscope is located in the School for material science studies. The School has its own fleet of ships to carry out marine minerals exploration studies world-wide.

Outstanding Local Facilities and Features
The City of Oxford combines the charm and safety of a small town with the sophistication of a larger city. USA Today named Oxford the "Thriving New South Arts Mecca" and one of the top six college towns in the nation. A national reputation brings major speakers, musicians, athletic events, and symposia to the city including the Faulkner and Yoknapatawpha Conference, the Oxford Conference for the Book, and the International Conference on Elvis Presley. Memphis is only one-hour's drive by interstate highway.

	$268 (credit hour)
	$268 (credit hour)
	750
	1 : 8
	42%
	Yes
	Yes
	1 April 1997
	Yes
	Yes
	1,000 acre resid. campus
	$70
	15 August 1997
	70 miles
	28 hours
	864 miles

New Mexico Institute of Mining and Technology

New Mexico Institute
of Mining and Technology

Taught Courses
Courses in support of graduate programs in
- Biology
- Chemistry
- Computer Science
- Engineering Science (mechanics)
- Materials
- Metallurgical
- Mining
- Petroleum and Environmental (engineering)
- Geoscience (geology, geochemistry, geophysics, hydrology)
- Mathematics and physics (atmospherics and Astrophysics)

Research Programs
Active research programs in biology (molecular, microbial, ecology, physiology, biochemistry); chemistry (biochemistry, environmental, explosive, inorganic, organic); computer science; engineering (mechanics, materials, metallurgical, mining, petroleum, environmental); geoscience (geology, geochemistry, geophysics, hydrology); mathematics; physics (atmospheric, astrophysics).

In addition to research in the college division there are research programs at the New Mexico Bureau of Mines; Petroleum Research and Recovery Center; National Radio Corporation of America and Energetic Materials Research and Testing Center.

General Departmental Information
For Information about graduate programs, contact:
J A Smoake, PhD
Graduate Office
801 Leroy Place
Socorro
NM 87801 USA
Email address: *Graduate@nmt.edu*

Special Departmental Resources and Programs
Graduate research facilities are supported by a number of on-campus research groups such as the New Mexico Bureau of Mines and Mineral Resources; the Petroleum Recovery Research Center; the Geophysical Research Center (hydrology, climatology and seismology); and the Energetic Materials Research and Testing Center.

Special facilities include the Langmuir Laboratory for Atmospheric Research (for studies of lightning, atmospheric physics and chemistry, and air quality); the 30-inch Colgate telescope; and the EMRTC Field Laboratory for Explosives Research. The Very Large Array radio telescope and the Very Long Baseline Array, both facilities of the National Radio Astronomy Observatory, are headquartered on the campus. Co-operative research opportunities are available with Sandia National Laboratories and Kirtland Air Force Base in Albuquerque and with Los Alamos National Laboratory.

Modern computer and library facilities and a wide range of modern analytical equipment are available.

Outstanding Local Facilities and Features
Langmuir Laboratory in the Magdalena Mountains; VLA facility on the plains of St Augustins; New Mexico Bureau of Mines (Mineral Museum) Etscorn Campus Observatory Macey Theatre/Conference Center; 18 hole golf course; Bosque del Appache Wildlife Refuge; Festival of the Cranes; San Miguel Catholic Church; Owl Bar and Cafe; Val Verde Steak House; Sandia National Laboratories; Los Alamos National Laboratories; White Sand Missle Range; numerous fiestas during the year.

Contact
J A Smoake, PhD
New Mexico Institute of
Mining and Technology
801 Leroy Place
Socorro
New Mexico 87801
USA

Tel (+1) 505 835 5513
Fax (+1) 505 835 5476
Email Graduate@nmt.edu
WWW nmt.educ

$83 (credit hour)	
$341.55 (credit hour)	
1,300	
–	
58%	
Yes	
–	
1 June 1997	
–	
–	
–	
–	
–	
75 miles	
–	
150 km	

The City College and the Graduate School of The City University of New York
School of Engineering

Contact
Dr Gerard G Lowen
Executive Officer
Graduate Engineering
The City College of New York
New York
NY 10031
USA

Tel   (+1) 212 650 8030
Fax   (+1) 212 650 8029

Taught Courses

MASTERS OF SCIENCE

MASTER OF ENGINEERING

DOCTOR OF PHILOSOPHY
Degrees in Chemical, civil, electrical and mechanical engineering as well as Computer science
Master is 30 credits, PhD is 60 credits plus dissertation

Research Programs

CHEMICAL ENGINEERING
Multiphase fluid mechanics; Fluidization; Reaction engineering; Process control and simulation; etc.

CIVIL ENGINEERING
Structural and geotechnical; Transportation engineering; Environmental engineering; Water resources.

COMPUTER SCIENCE
Computer systems; Scientific computing; Graphics; Cryptography; Artificial intelligence; Neural networks; Distributed computing; etc.

ELECTRICAL ENGINEERING
Photonics; Signal processing; Communication and network engineering; Biomedical engineering; Control theory; etc.

MECHANICAL ENGINEERING
Fluid dynamics; Vibration and control; Composite materials; CAD/CAM; Machine dynamics; Biomedical Engineering

General Information
The average graduate engineering enrolment at:
–   The Master's level is 500
–   The PhD level it is 250
–   The undergraduate level it is 2,600
Total City College student population is 14,000

Outstanding Local Facilities and Features
The City College occupies a 35-acre modern campus in historic St Nicholas Heights, only a few blocks from the Hudson River. Some of its buildings are New York City landmarks. In the surrounding area, one can find every kind of ethnic food. The New York City subway, which is only minutes away, provides access to Shopping centres; Parks; Museums, and Exciting sport centers in the city.

$2,175 (semester)	
$3,800 (semester)	
–	
–	
–	
–	
–	
Apr & May/ Nov & Dec	
Extensive	
Extensive	
–	
–	
–	
–	
–	
–	

on the net **http://www.editionxii.co.uk**

Rensselaer Polytechnic Institute
School of Engineering

Taught Courses
AERONAUTICAL ENGINEERING, MENG, MS, DENG, PHD

BIOMEDICAL ENGINEERING, MENG, MS, DENG, PHD

CHEMICAL ENGINEERING, MENG, MS, DENG, PHD

CIVIL ENGINEERING, MENG, MS, DENG, PHD

COMPUTER AND SYSTEMS ENGINEERING, MENG, MS, DENG, PHD

DECISION SCIENCES AND ENGINEERING SYSTEMS, PHD

ELECTRICAL ENGINEERING, MENG, MS, DENG, PHD

ELECTRIC POWER ENGINEERING, MENG, MS, DENG, PHD

ENGINEERING PHYSICS, MS, PHD

ENGINEERING SCIENCE, MS, PHD

ENVIRONMENTAL ENGINEERING, MENG, MS, DENG, PHD

INDUSTRIAL AND MANAGEMENT ENGINEERING, MENG, MS

MANUFACTURING SYSTEMS ENGINEERING, MS

MATERIALS SCIENCE AND ENGINEERING, MENG, MS, DENG, PHD IN MATERIAL ENGINEERING

MECHANICAL ENGINEERING AND MECHANICS, MENG, MS, DENG, PHD

NUCLEAR ENGINEERING, MENG, MS, DENG, NUCLEAR ENGINEERING AND SCIENCE, PHD

OPERATIONS RESEARCH AND STATISTICS, MS

TRANSPORTATION ENGINEERING, MENG, MS, DENG, PHD

General Departmental Information
The faculty is characterized by achievers and doers. Of the approximately 135 tenure-track faculty, 98% already hold their PhDs and in 1995–96, they were awarded over $41 million in sponsored research.

Rensselaers's optimum size blends individual program flexibility with broad based research opportunities, Rensselaer is known for its leadership and innovation in engineering education. US News & World Report ranks our graduate engineering programs in the first tier (top 25) overall, and in the top 10 in reputation by practicing engineers.

Outstanding Achievements of the Academic Staff to Date
Many well known publications: *Fundamentals of Physics*, by Robert Resnick, Halliday and Resnick, *Elementary Differential Equations and Boundary Value Problems* by W E Boyce and R C DiPrima, *Behavior in Organizations* by Robert Baron, *Oriental Philosophies*, by John Koller, and from Schaum's Outline Series, *Theory and Problems of Thermodynamics*, by Michael Abbott and Hendrick Van Ness.

Special Department Resources and Programs
Center for Integrated Electronics and Electronic Manufacturing (CIEEM); Center for Multiphase Research (CMR); Center for Services Sector Research and Education (CSSRE); Scientific Computation Research Center (SCOREC); Center for Composite Materials and Structures (CCMS); Lighting Research Center; Fresh Water Institute (FWI); Center for Image Processing Research (CIPR); Center for Infrastructure and Transportation Studies (CITS); New York State Center for Advanced Technology in Automation; Robotics; and Manufacturing (CAT); New York State Center for Polymer Synthesis (CPS); and Center for Entrepreneurship of New Technological Ventures (CENTV).

Contact
Vicki Lynn
Assistant Dean of Engineering
Rensselaer Polytechnic Institute
School of Engineering
Room 3004
Troy
New York 12188-3590
USA

Tel (+1) 518 276 6203
Fax (+1) 518 276 8788
Email lynnv@rpi.edu
WWW
http://www.rpi.edu/dept/grad~admissions

	$570 per credit hour
	$570 per credit hour
	1800 (ft)/ 600 (pt)
	1 : 2
	35%
	Yes
	–
	1 February 1997
	Yes
	Yes
	Suburban Campus
	Varies
	25 August 1997
	10 miles
	10 miles
	10 miles

UNIVERSITY OF
ROCHESTER

University of Rochester
School of Engineering and Applied Sciences

Contact

Duncan T Moore, Dean
School of Engineering and
Applied Sciences
University of Rochester
Rochester NY 14627
United States

Tel   (+1) 716 275 4151
Fax   (+1) 716 461 4735
Email
graddean@seas.rochester.edu
gradinfo@che.rochester.edu
gradinfo@me.rochester.edu
gradinfo@ee.rochester.edu
gradinfo@optics.rochester.edu
WWW
http://www.rochester.edu/

Taught Courses

–   Chemical Engineering: Major areas of study are bioengineering, interfacial phenomena and surface science, reaction engineering, system dynamics, and transport phenomena and separations
–   Electrical Engineering: General areas of study are biomedical engineering and medical ultrasound, solid devices and superconductivity, electromechanics of particles, computer engineering and applications, robotics, microelectronics, and VLSI design
–   Mechanical Engineering: Major areas of study are in biomechanics, laser fusion, and materials science
–   Optics: Instruction is available in optical instrumentation and design, quantum optics, laser engineering, signal processing, guided wave optics, non-linear optics, and optical materials

Research Programs

–   Chemical Engineering: Complex chemical networks, biocatalysis, phase transfer reactions, electrochemical sensors, molecular transport, plasma etching, supercritical phenomena, reactions, in porous media, polymerization, nucleation phenomena, fermentation, computer control, and optical materials
–   Electrical Engineering: Sound propagnation in tissue with applications to diagnosis, therapy and surgery; biological effects of both ultrasound and electromagnetic fields, including electrofusion; medical imaging, digital image and image sequence processing, pattern recognition; studies of fast relaxation processes in semiconductors and in superconductors by use of picosecond laser and microwave pulse; studies of fast kinematic strategies for intelligent robot control; analysis and design of computer-based design tools for enhancing productivity of analog and digital circuit designers, semiconductor device modelling, biomedical instrumentation, and multi-access protocols for computer communications
–   Mechanical Engineering: Energy and plasma physics, including inertial confinement nuclear fusion; applied mechanics, including flexible structures wave propagation, computer-aided analysis and design, astrophysical fluid dynamics, and experimental fluid mechanics; biomechanics, dealing with both ends of the cardiovascular system, biomedical applications for propagating waves, and the diagnosis and treatment of cancer; and materials science, emphasizing the relationship between microstructure and physical properties with special attention to mechanical properties
–   Optics: Optical instrumentation and design, quantum optics, laser engineering, signal processing, guided wave optics, non-linear optics, optical materials, and optical manufacturing. Well-equipped laboratories allow student research in gradient index optics, image processing, integrated optics, dielectric thin films, ultra-high resolution laser spectroscopy, and high-power laser physics.

General Information

The School of Engineering and Applied Sciences offers programs leading to MS and PhD degrees in Chemical Engineering, Electrical Engineering, Materials Science, Mechanical Engineering, and Optics. The MS degree requires a minimum of 30 semester hours of graduate credit and may be earned with or without a thesis. Thesis research can involve up to 12 hours of graduate credit. The MS option without a thesis may include up to 6 credits of independent study or project work and requires a comprehensive final examination.

Programs are open to both full-time and part-time students. A special part-time master's program designed to accelerate the time required to earn the degree is available for selected employees in local industries. The PhD degree is offered to prepare individuals for careers in research and teaching. The requirements include 90 semester hours of credit beyond the bachelor's degree and at least one academic year of full-time study in residence. A typical academic program is divided between course work and research credits to provide PhD candidates with a broad exposure to their fields of interest, the requisite training for mastery of their area of specialization, and experience in conducting scholarly research. PhD students must pass a preliminary examination and an oral qualifying examination, and must present and defend an original thesis that contributes to knowledge in the field.

Special Departmental Programs

Biomedical Engineering: This is an interdisciplinary program which includes faculty from all four departments within the School of Engineering and Applied Sciences, plus faculty from the School of Medicine and Dentistry. Faculty in all these departments are engaged in research that has biomedical applications, and they teach courses covering a variety of biomedical engineering principles and applications. At the time of this printing, graduate students have the option of enrolling in any of the four engineering departments and choosing a research speciality in biomedical engineering. Beginning in 1997, we anticipate that students will be able to enroll in an independent program leading to a master's or doctoral degree in biomedical engineering.

Materials Science: This program encompasses a variety of research activities in which advanced materials are tailored for specific uses. It allows students to be flexible in that they associate with the department closest to their interest in the field of materials science.

The Department of Mechanical Engineering has administrative responsibility for MS and PhD degrees.

Special Departmental Resources

Rochester Center for Biomedical Ultrasound: This resource unites more than 50 professionals from 18 departments of the University of Rochester, Rochester General Hospital, and the monographers' program at the Rochester Institute of Technology in pursuit of various aspects of physics, chemistry, engineering, biology, and medicine which make up the science and practice of biomedical ultrasound. The ongoing research of the Center provides a wide range of opportunities for thesis research.

Laboratory for Laser Energetics: This is a unique national resource for advanced research and education in technology related to the application of high-power lasers. Graduate students join the Laboratory by registering in one of the graduate degree programs within the University.

Center for Optics Manufacturing: This is the nucleus of a nationally supported research and development alliance that is modernizing the precision optics industrial base. The Center is a unique industry, university, and government collaboration that is redefining technology development and creating real-world solutions that increase industry manufacturing capability.

Center for Electronic Imaging Systems: This is a collaborative effort of the University of Rochester, the Rochester Institute of Technology, and industrial leaders in electronic imaging. Its efforts are based on the strategic identification of the basic research and technologies that support the future expansion of electronic imaging systems in the commercial, consumer, and governmental marketplaces.

📋	$615/credit hour (96-97)
💲	$615/credit hour (96-97)
👥	270 (ft/pt)
	Approx. 1 : 5
%	Approx. 60%
	–
	–
	15 December/1 Feb 97
	Yes
	Yes
?	Centralized
🛏	US $6,930 / year
	2 Sept 1997/14 Jan 1998
✈	2 miles
	11 hours (rail)
🚌	360 miles

on the net   http://www.editionxii.co.uk

Santa Clara University
Graduate School of Engineering

Taught Courses
Santa Clara University offers Master of Science, Engineer, and Doctor of Philosophy degree programs. MS degrees are offered in applied mathematics, civil engineering, computer engineering, electrical engineering, engineering management, and mechanical engineering. The school grants the PhD in electrical engineering, computer engineering, and mechanical engineering and the Engineering Degree in electrical and mechanical engineering. Graduate engineering courses are primarily offered one day a week from 7 to 9 am, with additional courses offered in the evening.

Research Programs
The school's research facilities include laboratories in the areas: object-oriented database systems, distributed computer systems, biomechanics, robotics and controls, circuits, image processing, digital systems, geotechnical testing, materials, water quality engineering, semiconductor fabrication, thermofluids, dynamics and control, and mechanical instrumentation. Computer facilities include a DEC Alpha; more than seventy-five state of the art engineering workstations, most of which are manufactured by Hewlett-Packard and Sun Microsystems; and many personal computers. Most computers are connected to the campus-wide Ethernet and to the Internet for international access.

The Institute for Information Storage Technology, organized under the auspices of the university, serves as a key technical resource to the information storage industry and provides outstanding educational opportunities for students in the relevant areas of data storage.

General Departmental Information
Santa Clara, the first institution to offer classes in higher learning in California, is a private Jesuit university. Graduate programs are conducted in the Leavey School of Business and Administration, the Division of Counseling Psychology and Education, Pastoral Ministries, and the School of Law, as well as in the School of Engineering.

The SCU student body consists of 3,800 undergraduates and 4,000 graduate and law students. There are about 600 undergraduates in the School of Engineering. Approximately 1,100 graduate students are in the graduate engineering programs. About two-thirds of the graduate students are employed in Silicon Valley and attend SCU on a part-time basis.

Special Departmental Resources and Programs
Santa Clara University's location offers a unique opportunity to acquire relevant work experience while working toward a degree. More than 100 Silicon Valley companies employ SCU co-op students either during alternative terms or on a part-time basis. Selection is limited to students who have completed 12 units with a GPA of 3.0 or better. International students must have completed one year of US residence prior to beginning a co-op. For further information, contact the Cooperative Education and Internship Office located in the Career Services building in Benson Center, (+1) 408 554 4470.

Outstanding Local Facilities and Features
Santa Clara University is located in the heart of Silicon Valley. The campus is situated in the midst of the nation's great concentrations of high-tech industry and professional and scientific activity. Santa Clara is 46 miles from San Francisco and only a few miles away from the San Jose International Airport. In the opposite direction, the Pacific beaches of Santa Cruz are about 30 minutes away; the world famous Monterey Peninsula and Carmel are two hours away. Santa Clara has a moderate Mediterranean climate. The average maximum temperature is 71 degrees and the average minimum, 42 degrees. The sun shines an average of 293 days per year, and the average annual rainfall is 15 inches.

Contact
Margaret P Seever
Santa Clara University
Graduate School of Engineering
500 El Camino Real
Santa Clara
CA 95053
USA

Tel   (+1) 408 554 4313
Fax   (+1) 408 554 5474
Email engr-grad@scu.edu
WWW
http://www.eng.scu.edu

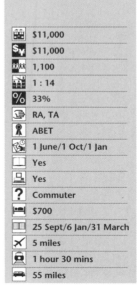

$11,000
$11,000
1,100
1 : 14
33%
RA, TA
ABET
1 June/1 Oct/1 Jan
Yes
Yes
Commuter
$700
25 Sept/6 Jan/31 March
5 miles
1 hour 30 mins
55 miles

UNITED STATES

University of South Alabama
College of Engineering

UNIVERSITY OF
SOUTH ALABAMA

Contact
Joseph D Sheehan
Tel (+1) 334 460 6050
Fax (+1) 334 460 7023
Email
jsheehan@jaguar1.usouthal.edu

David T Hayhurst, Dean
Tel (+1) 334 460 6140
Fax (+1) 334 460 6343
Email
dhayhurs@jaguar1.usouthal.edu

University of South Alabama
307 University Boulevard
Mobile
Alabama
USA 36688

Taught Courses
Graduate courses are taught in each of the five degree program areas: Chemical, Civil and Environmental, Computer, Electrical and Mechanical Engineering. Two plans of study leading to the MS degree in Engineering are offered. The thesis option requires a minimum of 48 quarter hours of which a minimum of eight quarter hours is thesis research. The project option requires a minimum of 52 quarter hours of which a minimum of four hours is project work. Industrial internships are possible when appropriate.

Research Programs
The University of South Alabama College of Engineering offers master of science degrees in Chemical, Civil and Environmental (proposed), Electrical and Computer, Mechanical Engineering. Doctoral degrees are offered in cooperation with other Alabama institutions. The college focuses its resources and efforts in areas of unique strength. By program, these areas are:

CHEMICAL ENGINEERING
Process design and simulation, estimation of thermodynamic properties, process control and molecular sieve synthesis and adsorption (email: thuddles@jaguar1.usouthal.edu).

CIVIL AND ENVIRONMENTAL ENGINEERING
Coastal engineering, constructed wetlands, drinking water treatment, innovative wastewater treatment, river basin modeling, lunar construction and speciality concretes (email: jolsen@jaguar1.usouthal.edu).

ELECTRICAL ENGINEERING
Optical computing, smart optical RAM and holography laser processing of electronic materials, giant magnetoresistive sensors, spin-valves and GMRAM solid state memory, high-energy power switching, microwave technology, wireless communication, acoustics and bioelectronics (email: mparker@jaguar1.usouthal.edu).

MECHANICAL ENGINEERING
Computational fluid mechanics, modeling and simulation in material processing, aerodynamics in diffusers and combustors, heat transfer and fluid mechanics, biomechanics and orthopaedic biomechanics, pedagogical research and teaching psychology (email: aengin@jaguar1.usouthal.edu).

Further details of faculty research interests and copies of recent publications can be obtained directly by contacting the specific departments. The University of South Alabama's homepage (http://www.usouthal.edu/usa/departms.html) has details concerning teaching programs, faculty and their research activities.

General College Information
The University of South Alabama is a comprehensive university and includes Colleges of Medicine, Arts and Science, Allied Health Professions, Business and Management Studies, Education, Engineering and Nursing. The College of Engineering has four separate departments: namely, Chemical, Civil and Environmental, Electrical and Computer, and Mechanical Engineering. The college has over thirty full-time faculty, approximately 1,000 undergraduate and 100 graduate students. All faculty in the College of Engineering hold doctoral degrees. The college has a large international student population. All programs are accredited by ABET. Each department has extensive computational facilities and direct access to a Cray supercomputer. There are well-equipped machine and electronic workshops and a full range of analytic facilities (including XRD and electron microscopy). All students have access to the University library's extensive collection. Full time graduate students typically receive assistantships.

General Information
Mobile is a thriving port city located on the Gulf of Mexico with a rich international heritage and a population of slightly over 500,000. The Gulf coast beaches, as well as New Orleans and Atlanta, are easily accessible by automobile. Mobile offers numerous museums and historic sites, as well as fine and performing arts; ballet, opera, theater and symphony.

The climate in Mobile is semitropical making outdoor activities enjoyable year round. The summers are warm and the winters mild, with average temperatures ranging from 51.4F in January to 81.8F in August. The Mobile area has 219 days of sunshine every year (60% of the year).

	–
$	$7,916 AY
	12,254 (Uni)
	–
%	6.3%
	No
	No
	15 June 1997
	Yes
	Yes
?	–
	$125–$150
	20 September 1997
✈	5 miles (8 km)
	10 miles (16 km)
	8 miles (13 km)

South Dakota School of Mines and Technology
Department of Civil and Environmental Engineering

MS IN CIVIL ENGINEERING
PhD IN ATMOSPHERIC, ENVIRONMENTAL AND WATER RESOURCES
The Department of Civil and Environmental Engineering offers the MS in Civil Engineering degree in Advanced Materials, Environmental Engineering, Geotechnical Engineering, Structural Engineering, and Water Resources Engineering. Both thesis and non-thesis programs are possible. Individualized, interdisciplinary programs of study are possible. The PhD in Atmospheric, Environmental, and Water Resources is offered jointly with South Dakota State University.

Faculty Speciality Areas and Research Programs
ENVIRONMENTAL ENGINEERING
Water and wastewater treatment process kinetics; municipal and industrial wastewater treatment; environmental chemistry; environmental contaminant fate and transport; adsorption processes; diffusion of contaminants across low permeability barriers; groundwater–surface water interactions; watershed water quality indicators; water quality field measurement methods; solid waste.

GEOTECHNICAL ENGINEERING
Theoretical soil mechanics; numerical modeling; constitutive relationships; soil reinforcement; polystyrene block applications; anchor-mooring system analysis; fiber reinforced soils; limiting equilibrium analysis; swelling of soils; properties of shales and overconsolidated soils; foundation engineering, exploration; soil dynamics.

STRUCTURAL ENGINEERING AND ADVANCED MATERIALS
Composite materials and structures; expert and knowledge-based systems; finite element analysis; plastic analysis and design of plane and space structures; shear wall structures; cable suspended structures; plastic analysis of prestressed concrete structures; fracture mechanics.

CONCRETE TECHNOLOGY
High performance; fiber reinforced; latex modified; high volume fly ash; ultra-high strength; high-durability; alkali-aggregate reactions; high performance concrete from recycled and waste materials; high performance concrete optimization; fast track paving; constitutive relations; fatigue and impact effects.

WATER RESOURCES ENGINEERING
Hydrology; rainfall runoff modeling; sediment transport; floods; hydraulics; water quality assessment; statistical analysis of water quality data; engineering hydrology; storm water management.

General Information
South Dakota School of Mines and Technology is a state supported university with approximately 2,200 students and 125 full-time faculty and researchers. The Department of Civil and Environmental Engineering has approximately 250 undergraduate students, 30 graduate students and 12 full-time faculty.

Special Departmental Resources and Programs
The Department of Civil and Environmental Engineering has over 450 square meters of materials testing, geotechnical engineering, environmental engineering, and structural engineering laboratories. Analytical and testing equipment includes fully instrumented MTS machines, a gas chromatograph, a high-performance liquid chromatograph, an atomic absorption spectrometer, a UV/visible spectrophotometer, and a total carbon analyzer. Geographic information systems computer work stations are available. Available on campus are an inductively coupled plasma spectroscope, transmission and scanning electron microscopes, an X-ray diffraction spectroscope, an electron microprobe and ion chromatograph. Computer facilities include five IBM RS-6000 computers and 125 clustered 486 and Pentium computers.

Research or teaching assistantships are available to qualified students. Teaching and research assistantships are usually eligible for reduced tuition. Full-time nine-month stipends range from $7,500 to $9,000. South Dakota Water Resources Fellowships are available for students who focus research on a State water resource concern. Students may participate in Cooperative Education and Cooperative Research Programs with the district Water Resources Division, US Geological Survey.

Outstanding Local Facilities and Features
Rapid City, South Dakota, population about 60,000, is located on the eastern edge of the Black Hills of South Dakota, in the heart of the American Old West. The Black Hills are gently rolling pine-clad hills and jagged spires towering 7,000-plus feet. Within easy driving distance of Rapid City are Mount Rushmore National Monument; Custer State Park, the nation's second largest state park; Harney Peak, highest point in the US east of the Rocky Mountains; Badlands National Park; Wind Cave National Park, and Jewel Cave National Monumont. Opportunities exist for downhill and cross-country skiing, camping, hiking, technical climbing, bicycling, swimming, hunting, fishing, spelunking, sailing, and boating. The climate is semi-arid and is moderated in the winter by the Black Hills, resulting in an excellent living environment (residents refer to the area as "the Banana Belt").

Contact
Dr Wendell H Hovey
Department of Civil and Environmental Engineering
South Dakota School of Mines and Technology
501 East Saint Joseph Street
Rapid City
South Dakota 57701-3995
USA

Tel   (+1) 605 394 2444
Fax   (+1) 605 394 5171
EMail whovey@silver.sdsmt.edu
WWW
http://www.sdsmt.edu

$2,500	
$5,500 (tuition & fees)	
2,200	
1 : 18	
9%	
–	
–	
Continuous	
Yes	
Residential/commuter	
$180	
8 January 1997	
3 September 1997	
10 miles	
No service	
1,600 miles (Washington)	

WASHINGTON·UNIVERSITY·IN·ST·LOUIS
School of Engineering & Applied Science

Washington University
School of Engineering and Applied Science

Contact
Christopher I Byrnes
Dean
School of Engineering
and Applied Science
Campus Box 1163
Washington University
One Brookings Drive
St Louis
MO
USA

Tel   (+1) 314 935 6166
Fax   (+1) 314 935 6949
Email
ChrisByrnes@seas.wustl.edu
WWW   www.cec.wustl.edu

Taught Courses
The School of Engineering and Applied Science, through its graduate and professional division, the Sever Institute of Technology, offers programs of instruction and research leading to the following degrees

MASTER OF SCIENCE
Biomedical Engineering; Chemical Engineering; Civil Engineering; Computer Science; Electrical Engineering; Engineering and Policy; Environmental Engineering; Management of Technology; Materials Science and Engineering; Mechanical Engineering; Systems Science and Mathematics; Technology and Human Affairs.

DOCTOR OF SCIENCE
Biomedical Engineering; Chemical Engineering; Civil Engineering; Computer Science; Electrical Engineering; Engineering and Policy; Environmental Engineering; Materials Science and Engineering; Mechanical Engineering; Systems Science and Mathematics.

PROFESSIONAL MASTERS
Construction Engineering; Construction Management; Control Engineering; Engineering Management; Information Management; Manufacturing Engineering; Structural Engineering; Technology Transfer in Electrical Engineering; Telecommunications Management.

Subject Areas of Engineering Research
Adaptive control; Aerospace systems; Air Pollution; Artificial intelligence; Astronomical imaging; Automation; Biomechanics; Biomedical engineering; Chemical reaction engineering; Communication and distributed computing technologies; Composite materials technology; Computational mechanics; Computer-aided drug design; Computer-aided process design; Computer-aided software engineering; Construction materials; Control theory and applications; Digital hearing aid design; Discrete-event analysis of digital systems; DNA sequence analysis; Dynamical systems; Earthquake engineering; Electrical power system operations; Electronics packaging; Environmental engineering; Groundwater hydrology; High-speed communication systems using ATM technology; Industrial catalysis; Information theory applications; Intelligent vehicle systems; International project management; Magnetic information systems; Management information systems development; Manufacturing systems; Mathematical modeling; Medical imaging; Networking; Nonlinear and linear systems; Optimal control and expert systems; Optoelectronics and photonics; Organizational dynamics of technology assimilation; Parallel processing; Plasma processing; Polymer science and engineering; Radar and sonar systems; Research management; Rheology; Robotics; Synthesis and processing of high performance materials; Transport phenomena.

General Information
The School of Engineering and Applied Science has seven academic departments (chemical engineering, civil engineering, computer science, electrical engineering, engineering and policy, mechanical engineering, and systems science and mathematics) and six interdisciplinary programs (biomedical engineering, computer engineering, environmental engineering, imaging science and engineering, materials science and engineering, and networking and communications). There are 1,000 undergraduate students and 800 graduate students, representing over 50 countries.

Outstanding Achievements of the Academic Staff to Date
Virtually all the 86 full-time faculty members of the School of Engineering and Applied Science hold doctorates. They include 53 editors of major scholarly publications, 27 Fellows in professional associations, three National Academy members, and seven National Young Investigators. Engineering faculty members and their students annually participate in over $20 million in externally funded research, divided about equally between public and private sponsors. In addition, over 100 engineers and scientists throughout the region are involved as affiliate faculty members, bringing a variety of advanced technical specialities into the school's classrooms and laboratories.

Special Departmental Resources and Programs
Applied Research Laboratory; Army Research Office National Center for Imaging Science; Center for Air Pollution Impact and Trend Analysis; Center for the Application of Information Technology; Center for Computational Biology; Center for Computational Mechanics; Center for Engineering Computing; Center for Optimization and Semantic Control; Center for Robotics and Automation; Center for Technology Assessment and Policy; Chemical Reaction Engineering Laboratory; Computer and Communications Research Center; Construction Materials and Management Center; Electric Power Research Institute Community Environmental Center; Electronic Systems and Signals Research Laboratory; Engineering Computing Laboratory; Fred Gasche Laboratory for Micostructured Materials Technologies; Jens Environmental Engineering Laboratory; Laboratory for Computational Science; Microelectronic Systems Research Laboratory; Materials Research Laboratory; Pen Laboratory; Technical Writing Center.

🏛	$9,996 (sem, 96–97)
💲¥	$9,996 (sem, 96–97)
👥	800
🏫	–
%	25%
🎓	Yes
🧑	–
📅	15 February 1997
▢	Yes
💻	Yes
?	Suburban
🛏	$200
▥	26 Aug/13 Jan/19 May
✈	10 miles
🚗	2 hours 30 mins
🚆	125 miles

on the net   http://www.editionxii.co.uk

Colorado School of Mines
Department of Geology and Geological Engineering

Taught Courses
Mineralogy; stratigraphy; structure; petrology; field methods; mineral deposits; petroleum geology; groundwater engineering; geotechnics; geomorphology; hydrology; groundwater modeling; oceanography; sedimentology; hydrogeology; mining geology; mineral exploration; mineralogy; clay characterization; remote sensing; geological data analysis; geotechnical aspects of waste disposal; carbonate and clastic sedimentology; petroleum reservoir characterization; geographic information systems; geochemistry (hard rock, low temperature, and environmental).

Research Programs
Exploration Geochemistry; Economic Geology; Structure; Geological Data Analysis; Sedimentology; Stratigraphy; Tectonics; Petroleum Geology; Formation Evaluation; Reservoir Characterization; Aqueous Geochemistry; Precious and Base Metal and Uranium Ore Deposits; Foundation Problems; Engineering Geology; Geotechnics; Slope Stability; Assessment and Mitigation of Geologic Hazards; Geotechnical Aspects of Waste Disposal; Metamorphic and Igneous Petrology; Paleoclimatology; Diagenesis; Remote Sensing; Aquifer Heterogeneities; Groundwater Flow and Transport Modelling; Basin Analysis; Geographic Information Systems; Continental Rifting.

General Departmental Information
The mission of our department is to provide leadership in undergraduate and graduate education and to develop and maintain strong, innovative applied research programs in hydrogeology, engineering and environmental geology, geochemistry, and both minerals and energy resource exploration and production with a foundation of basic geosciences. The department proactively pursues interdisciplinary endeavours, working cooperatively and productively with other departments and institutes on the Colorado School of Mines campus as well as with national and international societies, government agencies, academic institutions, and private corporations.

Outstanding Achievements of the Academic Staff to Date
- Roger M Slatt: AAPG Distinguished Service Award
- Robert J Weimer: (emeritus) AAPG Distinguished Educator Award, National Academy of Engineering
- John D Haun: (emeritus) AGI Heroy Award for Distinguished Service: AAPG Sidney Powers Memorial Award
- Samuel B Romberger: Pennsylvania State University Centennial Fellow of the College of Earth and Mineral Sciences
- Neil F Hurley: past AAPG Distinguished Lecturer
- Wendy J Harrison: past AAPG Distinguished Lecturer
- Murray W Hitzman: GSA Congressional Fellow, White House Office of Science and Technology Policy: AAAS/Sloan Fellow
- John E Warme: SEPM Honorary membership

Special Departmental Resources and Programs
- Petroleum Technology Transfer Council: assisting the petroleum industry with dissemination of petroleum exploration information and software technology
- International Groundwater Modelling Center: Research into, and courses concerning, groundwater modelling software
- Institute for Energy Resource Studies, which includes the US Potential Gas Agency

Outstanding Local Facilities and Features
Golden, Colorado is 15 miles west of Denver and combines a warm Western small-town atmosphere with establishments such as the National Earthquake Center, National Renewable Energy Laboratory, and Adolph Coors Brewing Company. The Denver metropolitan area provides the Colorado Symphony, the Denver Center for Performing Arts, Opera Colorado, museums, NBA basketball, NFL football, NHL hockey, major league baseball, zoo, and many other venues for music, theater and dance.

Contact
Roger M Slatt
Department of Geology and
Geological Engineering
Colorado School of Mines
Golden
CO 80401-1887
USA

Tel   (+1) 303 273 3822
Fax   (+1) 303 273 3859
Email rslatt@mines.edu
WWW   www.mines.edu

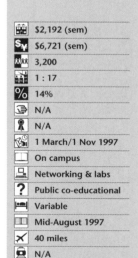

💰	$2,192 (sem)
💲	$6,721 (sem)
👥	3,200
👤	1 : 17
%	14%
📖	N/A
🧍	N/A
🗓	1 March/1 Nov 1997
🏫	On campus
💻	Networking & labs
?	Public co-educational
🍽	Variable
🎓	Mid-August 1997
✈	40 miles
	N/A
🚌	15 miles

Rensselaer Polytechnic Institute
School of Engineering

Contact
Vicki Lynn
Assistant Dean of Engineering
Rensselaer Polytechnic Institute
School of Engineering
Room 3004
Troy
New York 12188-3590
USA

Tel   (+1) 518 276 6203
Fax   (+1) 518 276 8788
Email lynnv@rpi.edu
WWW
http://www.rpi.edu/dept/
grad–admissions

🏛	$570 per credit hour
💲	$570 per credit hour
👥	1,800 (ft)/ 600 (pt)
🎓	1 : 2
%	35%
💳	Yes
👤	–
📅	1 February 1997
▢	Yes
💻	Yes
?	Suburban Campus
🛏	Varies
▥	25 August 1997
✈	10 miles
🚉	10 miles
🚌	10 miles

Taught Courses

AERONAUTICAL ENGINEERING, MENG, MS, DENG, PHD

BIOMEDICAL ENGINEERING, MENG, MS, DENG, PHD

CHEMICAL ENGINEERING, MENG, MS, DENG, PHD

CIVIL ENGINEERING, MENG, MS, DENG, PHD

COMPUTER AND SYSTEMS ENGINEERING, MENG, MS, DENG, PHD

DECISION SCIENCES AND ENGINEERING SYSTEMS, PHD

ELECTRICAL ENGINEERING, MENG, MS, DENG, PHD

ELECTRIC POWER ENGINEERING, MENG, MS, DENG, PHD

ENGINEERING PHYSICS, MS, PHD

ENGINEERING SCIENCE, MS, PHD

ENVIRONMENTAL ENGINEERING, MENG, MS, DENG, PHD

INDUSTRIAL AND MANAGEMENT ENGINEERING, MENG, MS

MANUFACTURING SYSTEMS ENGINEERING, MS

MATERIALS SCIENCE AND ENGINEERING, MENG, MS, DENG, PHD IN MATERIAL ENGINEERING

MECHANICAL ENGINEERING AND MECHANICS, MENG, MS, DENG, PHD

NUCLEAR ENGINEERING, MENG, MS, DENG, NUCLEAR ENGINEERING AND SCIENCE, PHD

OPERATIONS RESEARCH AND STATISTICS, MS

TRANSPORTATION ENGINEERING, MENG, MS, DENG, PHD

General Departmental Information
The faculty is characterized by achievers and doers. Of the approximately 135 tenure-track faculty, 98% already hold their PhDs and in 1995–96, they were awarded over $41 million in sponsored research.

Rensselaers's optimum size blends individual program flexibility with broad based research opportunities, Rensselaer is known for its leadership and innovation in engineering education. US News & World Report ranks our graduate engineering programs in the first tier (top 25) overall, and in the top 10 in reputation by practicing engineers.

Outstanding Achievements of the Academic Staff to Date
Many well known publications: *Fundamentals of Physics*, by Robert Resnick, Halliday and Resnick, *Elementary Differential Equations and Boundary Value Problems* by W E Boyce and R C DiPrima, *Behavior in Organizations* by Robert Baron, *Oriental Philosophies*, by John Koller, and from Schaum's Outline Series, *Theory and Problems of Thermodynamics*, by Michael Abbott and Hendrick Van Ness.

Special Department Resources and Programs
Center for Integrated Electronics and Electronic Manufacturing (CIEEM); Center for Multiphase Research (CMR); Center for Services Sector Research and Education (CSSRE); Scientific Computation Research Center (SCOREC); Center for Composite Materials and Structures (CCMS); Lighting Research Center; Fresh Water Institute (FWI); Center for Image Processing Research (CIPR); Center for Infrastructure and Transportation Studies (CITS); New York State Center for Advanced Technology in Automation; Robotics; and Manufacturing (CAT); New York State Center for Polymer Synthesis (CPS); and Center for Entrepreneurship of New Technological Ventures (CENTV).

Utah State University
Mechanical and Aerospace Engineering

Taught Courses
Finite Elements; Mechanical Vibrations; Composite Materials; Material Science; Thermal Systems Design; Orbital Mechanics; Advanced Orbital Mechanics; Spacecraft Attitude Control; Space Systems Design; Aerodynamics; Aircraft Stability and Control; Aeroelasticity; Advanced Dynamics; Continuom Mechanics; Elastic Theory; Plasticity Theory; Advanced Thermodynamics; Transport Phenomena, Heat Transfer; Viscous Fluid Flow; Optimal Control Theory.

Research Programs
Cryogenic Cooling Systems for Space-Based Instrumentation; Automated Systems; Spacecraft Altitude Control; Computational Fluid Dynamics; Space Structures; Composite Structures; Spacecraft Thermal Management; Space Systems; Astrodynamics; Aero-Dynamics.

General Departmental Information
Abet accredited undergraduate program. Aggressive graduate program focused on Advanced Treatment of Mechanical and Aerospace Engineering Fundamentals. USU is located in a very attractive dairy farming valley on the northern boundary of Utah.

Outstanding Achievements of the Academic Staff to Date
Faculty involvement in Space Research Programs has placed USU. in the forefront of US Space Research Program. USU. Faculty and students have flown more experiments on the Space Shuttle than any other university.

Special Departmental Resources and Programs
– Strong connection with USU. Space Research Program
– Strong program in Modern Control Systems
– Leading research institution in Cryogenic Cooling Programs

Outstanding Local Facilities and Features
– USU/AIAA Conference on small satellites (in the Fall)
– Outstanding winter sports
– Outstanding performing Arts programs

Contact
P Thomas Blotter
Mechanical and Aerospace
Engineering Department
Utah State University
Logan
Utah 84322 4130
USA

Tel   (+1) 801 797 2868
Fax   (+1) 801 797 2417
Email ptblotter@mae.usu.edu

🏫	$1,206
💲¥	$3,237
👥	25 (Dept)/20,000 (Ttl)
🎓	1 : 4.48
%	12%
📖	Yes
🔑	–
📅	1 January 1997
🖥	Yes
🖥	Yes
?	Small town
🛏	$150 per week
📅	September 1997
✈	83 miles
⚓	–
🚆	83 miles

Rensselaer Polytechnic Institute
School of Engineering

Contact
Vicki Lynn
Assistant Dean of Engineering
Rensselaer Polytechnic Institute
School of Engineering
Room 3004
Troy
New York 12188-3590
USA

Tel (+1) 518 276 6203
Fax (+1) 518 276 8788
Email lynnv@rpi.edu
WWW
http://www.rpi.edu/dept/
grad~admissions

Taught Courses

AERONAUTICAL ENGINEERING, MEnG, MS, DEnG, PhD

BIOMEDICAL ENGINEERING, MEnG, MS, DEnG, PhD

CHEMICAL ENGINEERING, MEnG, MS, DEnG, PhD

CIVIL ENGINEERING, MEnG, MS, DEnG, PhD

COMPUTER AND SYSTEMS ENGINEERING, MEnG, MS, DEnG, PhD

DECISION SCIENCES AND ENGINEERING SYSTEMS, PhD

ELECTRICAL ENGINEERING, MEnG, MS, DEnG, PhD

ELECTRIC POWER ENGINEERING, MEnG, MS, DEnG, PhD

ENGINEERING PHYSICS, MS, PhD

ENGINEERING SCIENCE, MS, PhD

ENVIRONMENTAL ENGINEERING, MEnG, MS, DEnG, PhD

INDUSTRIAL AND MANAGEMENT ENGINEERING, MEnG, MS

MANUFACTURING SYSTEMS ENGINEERING, MS

MATERIALS SCIENCE AND ENGINEERING, MEnG, MS, DEnG, PhD IN MATERIAL ENGINEERING

MECHANICAL ENGINEERING AND MECHANICS, MEnG, MS, DEnG, PhD

NUCLEAR ENGINEERING, MEnG, MS, DEnG, NUCLEAR ENGINEERING AND SCIENCE, PhD

OPERATIONS RESEARCH AND STATISTICS, MS

TRANSPORTATION ENGINEERING, MEnG, MS, DEnG, PhD

General Departmental Information
The faculty is characterized by achievers and doers. Of the approximately 135 tenure-track faculty, 98% already hold their PhDs and in 1995–96, they were awarded over $41 million in sponsored research.

Rensselaers's optimum size blends individual program flexibility with broad based research opportunities, Rensselaer is known for its leadership and innovation in engineering education. US News & World Report ranks our graduate engineering programs in the first tier (top 25) overall, and in the top 10 in reputation by practicing engineers.

Outstanding Achievements of the Academic Staff to Date
Many well known publications: *Fundamentals of Physics*, by Robert Resnick, Halliday and Resnick, *Elementary Differential Equations and Boundary Value Problems* by W E Boyce and R C DiPrima, *Behavior in Organizations* by Robert Baron, *Oriental Philosophies*, by John Koller, and from Schaum's Outline Series, *Theory and Problems of Thermodynamics*, by Michael Abbott and Hendrick Van Ness.

Special Department Resources and Programs
Center for Integrated Electronics and Electronic Manufacturing (CIEEM); Center for Multiphase Research (CMR); Center for Services Sector Research and Education (CSSRE); Scientific Computation Research Center (SCOREC); Center for Composite Materials and Structures (CCMS); Lighting Research Center; Fresh Water Institute (FWI); Center for Image Processing Research (CIPR); Center for Infrastructure and Transportation Studies (CITS); New York State Center for Advanced Technology in Automation; Robotics; and Manufacturing (CAT); New York State Center for Polymer Synthesis (CPS); and Center for Entrepreneurship of New Technological Ventures (CENTV).

🏛	$570 per credit hour
$	$570 per credit hour
👥	1800 (ft)/ 600 (pt)
🏫	1 : 2
%	35%
📖	Yes
🧍	–
🗓	1 February 1997
📋	Yes
🖥	Yes
?	Suburban Campus
🛏	Varies
📅	25 August 1997
✈	10 miles
🚗	10 miles
🚌	10 miles

University of Illinois at Urbana-Champaign

Taught Courses
Graduate courses in radiation transport; neutron transport; plasma transport; thermal energy transport; radiation interaction with matter; nuclear materials; health physics; radiobiology; nuclear reactor safety; non-linear methods in nuclear processes; monte carlo methods; fuzzy logic methods; nuclear instrumentation and measurements; direct energy conversion; and fission and fusion power systems. Wide variety available in other engineering disciplines.

Research Programs
- Fission and fusion research and power systems
- Health physics/neutron capture therapy/radiation biology
- Inertial electrostatic confinement devices for radiation source use
- Neutron scattering and reflectometry
- Neutron activation analysis
- Nuclear materials (corrosion, radiation damage, mechanical behaviour)
- Nuclear pumped lasers/energy conversion
- Plasma wall interactions, processes, and diagnostics
- Radioactive waste management, storage, and non-proliferation
- Thermal-hydraulics and reactor safety

General Departmental Information
The department offers BS, MS and PhD degrees in nuclear engineering.
It has a 1.5MW TRIGA nuclear reactor with pulsing capabilities to 6,000 MW peak power. There are 11 FTE faculty and 15 affiliated faculty, providing wide range of research topic selection.

Outstanding Achievements of the Academic Staff to Date
- Discovery and identification of several nuclear pumped lasers
- Development of plasma diagnostics, fluctuating static pressure measurement probes, etc.
- Inventions of inertial electrostatic confinement devices as neutron and x-ray sources
- Outstanding teaching (University and College Awards)

Special Departmental Resources and Programs
- Advanced analysis techniques using Lie groups and group invariant difference schemes
- Dense Plasma Focus Facility
- Fusion Studies Laboratory
- Nuclear Materials Laboratory
- Plasma Wall Interaction, Diagnostics and Processing Laboratory
- Thermal-hydraulics Laboratory
- TRIGA Nuclear Reactor/Neutron Activation Analysis/Small Angle Scattering Laboratory

Contact
Barclays Jones
Department of Nuclear Engineering
214 Nuclear Engineering Laboratory
103 South Goodwin Avenue
Urbana
IL 61801-2984
USA

Tel   (+1) 217 333 2295
Fax   (+1) 217 333 2906
Email nuclear@uiuc.edu

$1,950 per semester
$4,987 per semester
45 (Campus 8834)
–
34%/66%
Yes
–
June/October/March
Yes
Yes
Undergradte/Graduate
$79.00
3 Sept 1997/22 Jan1998
5 miles
3 hours to Chicago, IL
140 miles to Chicago, IL

Extended Entries

Universités de Bourgogne et de Franche-Comté

DEA Instrumentation et Informatique de l'Image (DEA 31)

Purpose
The goal of this DEA is to bring to the students a high level training in the Imaging area.

Schedule
The lectures are dispensed during the first semester (October to February).

During the second semester the students work on a specific research project in an University Laboratory or in a R/D company.

The details of the lectures program are:

Common Courses: 96 H
  Signal Processing
– Image Segmentation
– Markov Fields and Mathematics Morphology
– Vision Systems Architecture
– Image Synthesis
– Artificial Intelligence

Option 1: 96 H
Image acquisition and Imaging Systems) (*)
– CCD Cameras
– Real time Algorithms Implementation
– Radio-Isotopic Imaging Systems, Scanner X and IRM
– Infra Red Vision

Option 2: 96 H
Computing and Image Analysis (*)
– Image analysis
– Pattern Recognition and classification
– 3D Computing, Movements analysis
– Images data bases and Image Compression

(*) The students have the choice of one of the two options

Contact
Professeur Michel Paindavoine
Université de Bourgogne
6 Boulevard Gabriel
21 000 Dijon
France

Tél: (+33) 3 80 39 60 43
Fax: (+33) 3 80 39 60 43
Email: paindav@satie.u-bourgogne.fr

Ecole Normale Supérieure de Cachan

(Cinq km south from Paris – accomodation available on the campus)
The Ecole Normale Supérieure de Cachan, part of the French Grandes Écoles system, proposes an efficient research structure: 14 laboratories; 200 senior researchers and 250 doctorate students, which maintains collaborations with European and American major universities, research teams and industrial corporations.

For a student at graduate level, the research experience includes: One year for the DEA degree followed by three years for the Doctorate degree in 16 different topics.

Main fields:
Applied Mechanics, Civil and Mechanical Engineering
(three DEA)
– **Matériaux Avancés: Ingéniérie des Strucrures et des Enveloppes; Advanced comentitious composites; Modelling behaviour and computing under severe loadings**
– **Mécanique et Matériaux: From the macrostructure to the macrobehaviour of several materials**
– **Solides, Structures et Systèmes Mécaniques: Computational solid mechanics for composite materials and structures; Computer aided design**

Electronics, Electrotechnics, Robotics
(Four DEA)
– **Automatique et Traitement du Signal: Feedback control; Signal and image processing; Filtering wavelets**
– **Electronique, Capteur et Circuits Intégrés: Specific integrated circuits and microsystems; Sensors modelling and signal processing**
– **Génie Electrique: Electrical engineering; Power electronic; Microtechnology; Control of electrical drives**
– **Production Automatisés: Computer integrated manufacturing; Production systems modelling and engineering**

Other fields (Nine DEA)
Applied Mathematics and Computer Sciences
Applied Biology
Photophysics and Photochemistry
Risk, Information and Decision
Social Sciences and Politics

Contact
Professor Jacky Mazars
Deputy Director ENS Cachan
61 Avenue du Président Wilson
94 235 Cachan cedex
France

Tel (+33) 1 47 40 20 71
Fax (+33) 1 47 40 28 88
Email: mazars@ens-cachan.fr
WWW: http://www.ens-cachan.fr

Université de Lille I / Ecole Centrale de Lille

Electrotechnique

Formations Enseignées

DEA Génie Electrique

Interrupteurs à semi-conducteurs; contrôle des commutations; Structures de conversion; Commutation douce; Convertisseurs directs; Filtrage; Modélisation et commande rapprochée des convertisseurs; Conversion d'énergie; Modélisation des machines électriques; Analyse numérique des systèmes électrotechniques; CAO en électrotechnique; Stratégies de contrôle des machines électriques; Régulation des entraînements à vitesse variable; Machines spéciales; Matériaux magnétiques et aimants permanents.

Programmes de Recherche

Dans le cadre du Laboratoire d'Electrotechnique et d'Electronique de puissance (L2EP): les thèmes portent sur la conversion de l'énergie électrique; le traitement électronique de l'énergie électrique (structures de conversion et commande des processus électriques) et la modélisation numérique des systèmes électrotechniques.

Informations Générales

Objectifs et débouchés:

– **Compléter la formation en électrotechnique et électronique de puissance par des enseignements de haut niveau**
– **Initier à la recherche dans ces disciplines par un stage en laboratoire**
– **Préparer aux carrières de l'enseignement supérieur, aux emplois de cadres de recherche Prérequis: être titulaire de la Maîtrise EEA, ou** d'un diplôme équivalent.

Contact

Professeur R Bausière
Université des Sciences et Technologies de Lille – Lille I
Cité Scientifique
59 655 Villeneuve d'Ascq cedex
France

Tél: (+33) 3 20 43 42 35
Fax: (+33) 3 20 43 69 67
Email: bausiere@univ-lille1.Fr

Université de Limoges – Faculté des Sciences

DESS Communications Radiofréquences et Optiques

Taught Courses

Diplôme d'Etudes Supérieures Spécialisées (DESS)

Postgraduation in Electronics (Optical and Microwave Communications)

Lectures and practical training are given during six months (1 October to 31 March). Then, during four to six months, the students work with an industrial company in France or any other country.

Programmes and General Information

Computer aided design of microwave circuits (MDS, J-OMEGA, P-SPICE, Microwave Success, Explorer etc); Power amplifiers; Test and measurement of microwave and optical circuits; Optoelectronic and integrated optics; Electromagnetic wave propagation; Cellular telecommunications systems; Radar and Lidar – Active and passive microwave filters; non-linear active circuits; Solid state device modelling; Phase locked loops; Antennas; Electromagnetic compatibility.

Contact

Monsieur Michel Bridier
Faculté des Sciences
123 avenue Albert Thomas
87060 Limoges cedex
France
Tél: (33) 5 55 45 73 44
Fax: (33) 5 55 45 76 97
Email: nadine.bresson@ircom.unilim.fr

Secrétariat: Mme Nadine Bresson
Tél: (33) 5 55 45 72 50

Université de Haute Alsace – Ecole Nationale Supérieure des Industries Textiles de Mulhouse

DEA Génie des Processus et des Matériaux Textiles et Paratextiles

Le DEA est une formation de troisième cycle qui se déroule sur un an et permet la préparation d'un doctorat en Sciences de l'Ingénieur.

Il comporte des enseignements théoriques (140 heures) dans les domaines suivants:
- **Théorie des matériaux textiles à structure bloquée**
- **Mécanique des matériaux textiles à structure souple**
- **Dynamique des structures textiles**
- **Tribologie**
- **Rhéologie**
- **Matériaux textiles avancés**
- **Adhésion et multimatériaux**

Cet enseignement est complété par des conférences et séminaires ainsi que par un stage de recherche. Celui-ci peut être effectué dans l'un des laboratoires d'accueil de la formation doctorale:
- **Laboratoire de Physique et Mécanique Textiles (ENSITM)**
- **Laboratoire de Mathématiques (UHA-Mulhouse)**
- **Institut de Chimie des Surfaces et des Interfaces (CNRS – Mulhouse)**
- **Institut de Mécanique des Fuides (CNRS – Strasbourg)**
- **CIRAD – CA (Montpellier)**

Admission
- **Sur dossier**
- **Titulaires d'une maîtrise ou d'un diplôme d'ingénieur ou d'un diplôme étranger admis en équivalence dans les disciplines: mécanique; physique; chimie.**

Contact
Professeur D Dupuis
ENSITM
11 rue A Werner
68 093 Mulhouse cedex
France
Tel: (+33) 3 89 59 63 20
Fax: (+33) 3 89 59 63 39

Taganrog State University of Radio Engineering

Programs leading to Diploma in Engineering, Bachelor of Science and Master of Science Degrees

Radio Physics and Electronics; Industrial Electronics; Radio Engineering; Telecommunications, Broadcasting and Television; Vehicular Communications; Audio and Video Engineering; Consumer Electronics; Radio Systems Engineering; Computer Engineering, Automated Systems of Control and Data Processing; Computer Aided Design; Software Engineering; Standardisation and Certification; Hydroacoustics and Hydrophysics; Testing and Diagnostics; Applied Acoustics; Biomedical Engineering; Measurements and Instrumentation; Microelectronics and Semiconductor Devices; Electronic Devices and Apparatus; Design and Manufacturing in Radio and Computer Engineering; Environmental Engineering; Economics; State, Municipal and Business Administration; Management; Data Systems in Economics; Art and Culture.

Programs leading to Candidate of Science Degree and Doctor of Science Degree

Radio Physics; Acoustics, Electrical and Computer Engineering; Antennas and Microwaves; Radio Engineering Systems; Medical Instrumentation; Control Systems; CAD; Computer Science; Microelectronics; Education.

Research Programs

Multiprocessor computing systems; computer networks; intellectual systems; CAD system; medical diagnostic systems; sensors; SHF solid-state electronics; design and manufacturing of VLSI circuits; radio signal processing; hydroacoustic and ultrasonic devices; electrodynamic structures; circuits and signals modelling; information means for self-investigation, self-evaluation and self-education; problems of regional economic reforms; parallel computing systems with programmable architecture; distributed neuro engineering systems; multiprocessor and neuroprocessor information control systems; data protection; microelectronics multisensors systems; synergetic and control; reliability of devices and apparatus.

General Information

The University was established on 1952 in Taganrog, an industrial, commerce and port city on the South of Russia. The population of Taganrog is close to 300,000. A distinguishing features of the city are the low prices, mild climate and numerous parks, gardens and beaches.

Contact

A G Pilipenko
Head of International Co-operation Department
Taganrog State University of Radio Engineering
44, Nekrasovsky str
Taganrog GSP 17-A
Rostov Region
Russia 347928
Tel: (+7) 863 446 16 85
Fax: (+7) 863 444 18 76
Email: rector@trtu.rnd.su
http://WWW.tsuRE.RU/

University of Bradford

Department of Civil and Environmental Engineering

Research for MPhil and PhD

Applications are invited for research work in any of the following areas of research currently being carried out in the Department.

- **Structural Engineering (development of numerical and experimental methods and material models for the analysis and optimal design of engineering structures)**
- **Environmental Acoustics (development of experimental and numerical models for propagation of environmental noise abatement)**
- **Environmental Hydraulics (development and refinement of numerical models for predicting hydrodynamic, water quality and sediment transport processes relating to coastal, estuarine and inland waters)**
- **Geotechnics (research on effects of soil crushability and cyclic loads on foundations; degradable rock slopes, tunnels and geotechnical processes),**
- **Groundwater Protection and Restoration (research on physical, chemical and biochemical processes in subsurface for enhanced groundwater protection and restoration of contaminated land).**

MSc in Structural Engineering

The course is intended for those who wish to gain advanced level and detailed knowledge of topics of importance in structural engineering today.

The course follows the standard pattern with parallel Postgraduate Diploma and is only available full time.

Find more information on http://www.brad.ac.uk/acad/civeng/civhome.html

Contact
Dr V V Toropov
Research Co-ordinator
Department of Civil and Environmental Engineering
University of Bradford
Bradford
West Yorkshire
BD7 1DP
Tel: (+44) 1274 383 869
Fax: (+44) 1274 383 888
Email: v.v.toropov@bradford.ac.uk

University of Hull

Engineering Design and Manufacture

The Department was founded in 1979 and since that time its research reputation has developed rapidly.

In the 1992 research selectivity exercise it was one of very few UK Engineering departments to be awarded the highest rating.

Taught Courses

MSc in Advanced Manufacturing and Materials

A one-year full-time programme, combining formal teaching, written examinations and the submission of a short dissertation or project. A second class honours degree or better is normally required.

Research Programmes

MSc, MPhil and PhD degrees are offered by research in the following main research areas: acoustics and vibration; automatic monitoring and inspection techniques; design for economic manufacture; electronic materials; finite element analysis; mechanical properties of engineering ceramics; tribological coating technology; and wear mechanisms in ultra hard materials.

Degrees are normally of two years' duration (MSc/MPhil) or three years' (PhD).

General information

Founded in 1927, the University is situated on an attractive campus two and a half miles from the city centre of the historic city of Kingston upon Hull, and within easy reach of unspoiled countryside and coast.

Promoting Excellence in Education and Research

Contact
Dr R D James
Department of Engineering Design and Manufacture
School of Engineering and Computing
University of Hull
Hull HU6 7RX
United Kingdom

Tel: (+44) 1482 466222
Fax: (+44) 1482 466533

Southampton Institute

Systems Engineering Faculty

MSC Computing (Software Engineering)

A one calendar year full-time course focusing on the application of engineering techniques to the development of software, particularly in the real-time and embedded systems environment.

The course requires 30 weeks of taught study of five core units:
– **Embedded Systems**
– **Formal Methods**
– **Software Engineering Principles**
– **Software Testing and Validation**
– **System Development Methodologies**

Supported by two options from:
– **Adaptive Systems**
– **Communications Data Processing**
– **Data Communications and Networks**
– **Human Computer Interaction**
– **Object Oriented Systems**

Successful completion of this taught part of the course allows progression to a 15 week dissertation period of industry based or research study leading to the award of MSc Computing (Software Engineering).

Applications for the course are welcomed from candidates holding a good first degree in computing, engineering or science with experience of software development.

The Systems Engineering Faculty has excellent computing resources in support of this programme including private laboratory facilities dedicated to Postgraduate students. The Faculty offers a strong academic team of individual specialists many of whom are nationally known for their research or professional work on national committees.

Contact
Mark Udall
Systems Engineering Faculty
East Park Terrace
Southampton SO14 OYN
United Kingdom
Tel: (+44) 170 331 9322
Fax: (+44) 170 333 4441

University of Strathclyde

Civil Engineering

The Department of Civil Engineering of the University of Strathclyde has offered postgraduate MSc and Diploma courses for over thirty years. The courses have an enviable reputation for both the quality of their teaching and the relevance and practicality of the credits taught. Five courses are currently on offer:
– Water Engineering
– Hydraulics; Hydrology and Coastal Engineering
– Public Health and Environmental Control Engineering
– Highway Engineering
– Construction Management.

Each MSc course involves two instructional semesters in which passes have to be obtained in a minimum of ten credits, followed by a project. This is usually undertaken in collaboration with industry or the student's sponsor or employer.

The Department also offers opportunities for research at MPhil and PhD levels in all areas of Civil Engineering. The Department is particularly strong in geotechnics; water and environmental management; structures; and construction management. Excellent laboratories are backed up by the most advanced computer facilities and an extensive library.

Strathclyde University, celebrating its 200th Anniversary in 1996, is located in the centre of the City of Glasgow. Glasgow enjoys a world wide reputation for the welcome of its people, its diverse culture, sport, business and easy access to some of the finest scenery in the UK. With an international airport, Glasgow has fast connections to all parts of the world.

Accommodation is available within easy reach of the University for all postgraduates, including those with families, and special facilities exist to allow students from all backgrounds to feel "at home" in one of the UK's friendliest cities. Assistance can also be provided with English language, including pre-sessional instructional courses.

Contact
Dr John Riddell
Department of Civil Engineering
University of Strathclyde
Glasgow G4 0NG
United Kingdom
Tel: (+44) 141 552 4400
Fax: (+44) 141 553 2066

University of Missouri – Columbia

College of Engineering

Courses Taught
- Agricultural Engineering: MS; PhD
- Chemical Engineering: MS; PhD
- Civil Engineering: MS; PhD
- Computer Science: MS
- Computer Engineering: MS-EE; PhD-EE
- Electrical Engineering: MS; PhD
- Industrial Engineering: MS; PhD
- Mechanical & Aerospace Engineering: MS; PhD
- Nuclear Engineering: MS; PhD

Research Programs

All departments have state-of-the art research facilities. Excellent engineering and university library facilities provide the latest domestic and international journals specific to engineering and physical science research.

The College of Engineering computer center supports super-minicomputer; minicomputer, and microcomputer systems, including a VAX system with digital, HP, sun and SGI systems. The University also supports a state-of-the art virtual reality classroom teaching and developmental system.

The MU Research Reactor Center, a 10-megawatt facility has the highest power of any US university reactor. The Center also houses a 26,000 sq. ft. laboratory research facility.

Financial Assistance

Research and teaching assistantships are available to qualified students for the academic year. Fellowships are also available to students who do not qualify for assistantships. These fellowships are awarded for the academic year and may allow for waiver of the non-resident educational fee.

Application Information

Specific information pertaining to the application process and applicable deadlines may be obtained from the University of Missouri-Columbia Graduate Admissions Office or the College of Engineering.

Contact
E Robert Jones
Director Graduate Recruitment
University of Missouri
Columbia College of Engineering
W1025 EBE
Columbia
MO 65211-0001
USA
Tel: (+1) 573 882 4487 or (+1) 800 877 6312
Fax: (+1) 573 882 2490
Email: info@grad.missouri.edu
WWW: http://www.research.missouri.edu

Thayer School of Engineering

Dartmouth College

Thayer School of Engineering offers course curricula in areas of current and social importance, specifically biomedical engineering, biotechnology, environmental engineering, computer engineering and communication, electrical engineering, mechanical engineering, controls, materials science, and engineering management.

Thayer School currently offers over 140 courses spanning the many disciplines of engineering. Students are also encouraged to take courses in other areas of Dartmouth College, such as computer science, earth sciences, physics, chemistry, biology, business and medicine.

Contact
William Lotko
Director, MS, PhD Programs
Thayer School of Engineering
Dartmouth College
Hanover
NH 03755
USA
Tel: (+1) 603 646 3844
Fax: (+1) 603-646 3856
Email: thayer.admissions@darmouth.edu

Tufts University

Graduate Engineering

Four departments of engineering offer both the Master of Science and the PhD: Electrical Engineering and Computer Science, Civil and Environmental Engineering, Chemical Engineering and Mechanical Engineering. In addition, the Master of Science in Engineering Management is awarded through the Gordon Institute. Specialized areas of concentration include electro-optics, computer engineering, software engineering, hazardous materials management, water resources management, geotechnical engineering, environmental health, pollution prevention, thermal manufacturing and manufacturing engineering. Degree candidates are encouraged to pursue their projects in collaboration with local industries or the allied science departments. Scholarships and some teaching/research assistantships available.

Contact:

Graduate School
Tufts University
120 Packard Avenue
Medford, MA 02155
Tel: (+1) 617-627-3395
Fax: (+1) 617-627-3016
Email: GSAS@Infonet.Tufts.Edu

University of Utah

College of Engineering

Taught Courses

Departments
– **Bioengineering**
– **Chemical and Fuels Engineering**
– **Civil and Environmental Engineering**
– **Computer Science**
– **Electrical Engineering**
– **Materials Science and Engineering**
– **Mechanical Engineering**

Graduate Degrees Offered
– **Master of Engineering**
– **Master of Science**
– **Master of Philosophy**
– **Doctor of Philosophy**

General Information

With 2,800 students, the College of Engineering is known for teaching excellence and research strengths in bioengineering; robotics; computer graphics; combustion; advanced materials and polymers; microelectronics; electromagnetics; and natural resource development and exploration, among others. Research activities are typified by close interaction with industrial partners.

Outstanging Local Facilities and Features

The University of Utah is located on the northeastern edge of Salt Lake City in the foothills of the majestic Wasatch Mountains. Site of the 2002 Winter Olympics, the city is known internationally for it's excellent year-round outdoor recreational opportunities with access to several world-class ski resorts, and nine national parks.

Contact
Dean's Office
College of Engineering
Kennecott Building
University of Utah
Salt Lake City
UT 84112
USA

Tel: (+1) 801 581 6911
Fax: (+1) 801 581 8692
WWW: http://voyager.eng.utah.edu/